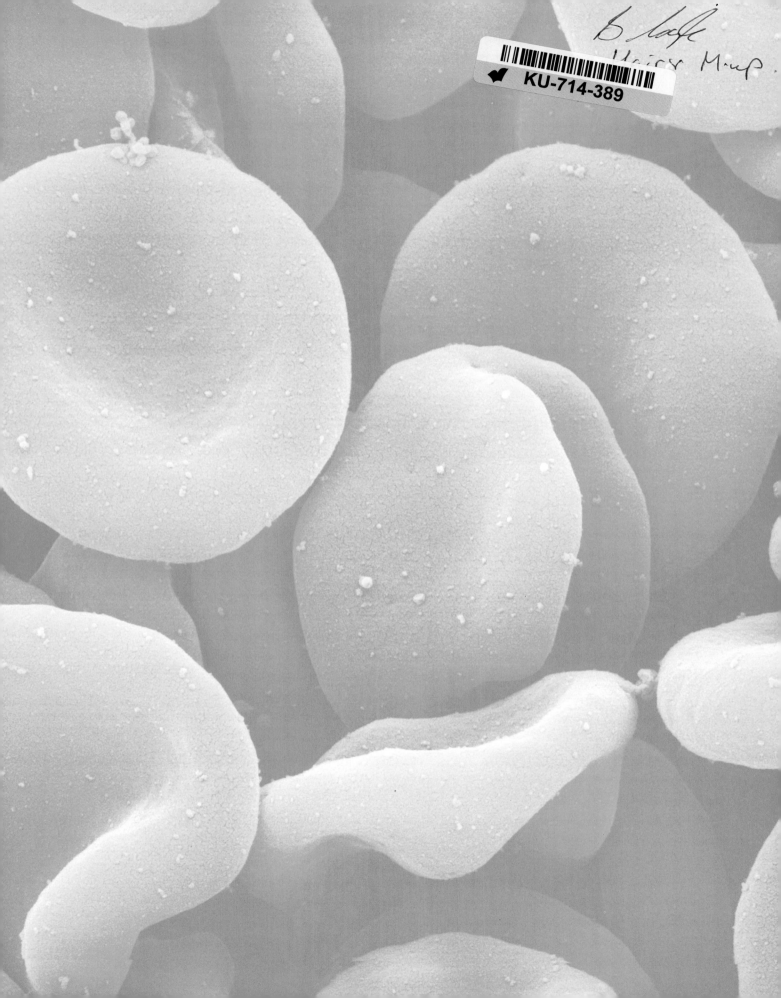

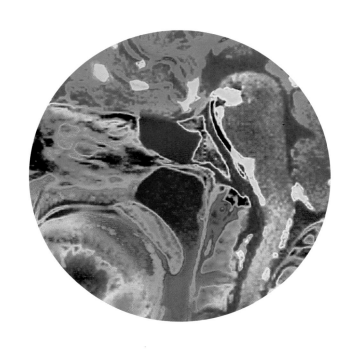

ENCYCLOPEDIA OF THE
HUMAN BODY

LONDON, NEW YORK, MUNICH,
MELBOURNE, and DELHI

for PAGE *One*:
Cairn House, Elgiva Lane,
Chesham, Bucks HP5 2JD

Creative director Bob Gordon
Managing editor Judith Hannam
Editors Naomi Mackay, Michael Spilling, Charlotte Stock
Art editor Robert Law
Designers Annie Hiew, Tim Stansfield
Special photography Andy Crawford
Digital artwork Anthony Duke, Mark Tattam, Peter Bull
Picture research John Farndon
With special thanks to Tania Fuller, Joe Gordon, Alex Jones,
Sophie Pateman, Adrian Phillips, and Henry Spilberg

for Dorling Kindersley:

Senior editors Selina Wood, Shaila Awan
Art editors Polly Appleton, Steven Laurie
Project editor Lucy Hurst
Managing art editor Clare Shedden
Managing editors Marie Greenwood, Andrew Macintyre
Digital artwork Robin Hunter
DTP design Siu Yin Ho
Production Kate Oliver
Picture research Sean Hunter
DK pictures Rose Horridge

Editorial consultant and author Richard Walker

Contributing authors David Burnie, Daniel Carter,
Phil Gates, Penny Preston, Frances Williams

First published in Great Britain in 2002
by Dorling Kindersley Limited
80 Strand, London
WC2R 0RL

A Penguin Company

2 4 6 8 10 9 7 5 3 1

A CIP catalogue record for this book
is available from the British Library.

ISBN 0-7513-3927-X

Colour reproduction by Colourscan, Singapore
Printed and bound in Slovakia by Neografia

See our complete catalogue at

www.dk.com

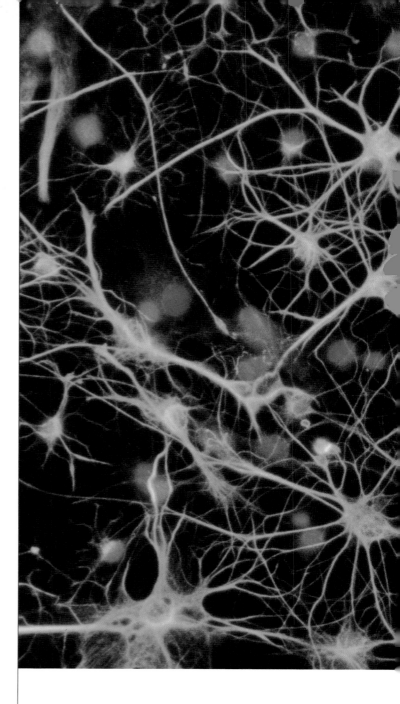

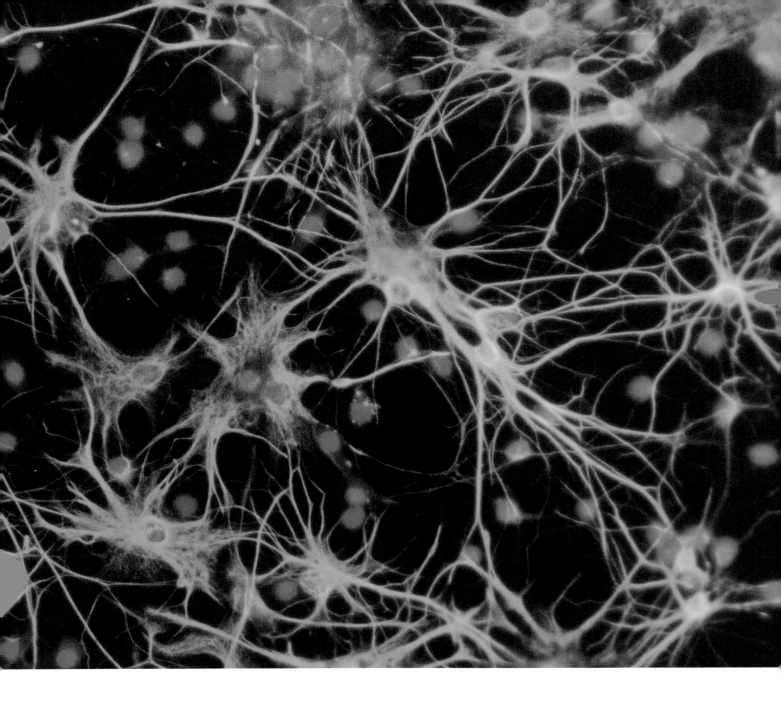

ENCYCLOPEDIA OF THE HUMAN BODY

Richard Walker

Contents

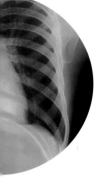
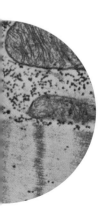

SUPPLY AND MAINTENANCE   126–209

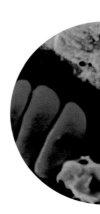

Contents

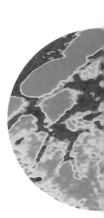

The following abbreviations have been
used for technical imaging techniques:

CT = Computed Tomography
LM = Light micrograph
NMR = Nuclear Magnetic Resonance Imaging
MRI = Magnetic Resonance Imaging
PET = Positron Emission Tomography
SEM = Scanning Electron Micrograph
TEM = Transmission Electron Micrograph

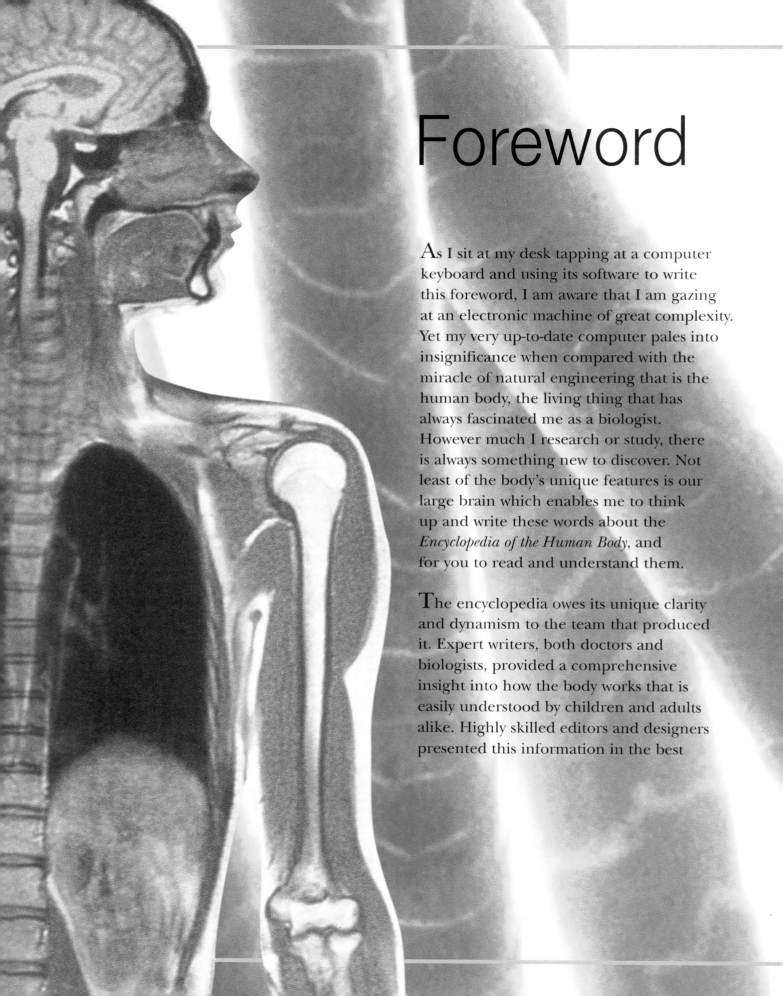

Foreword

As I sit at my desk tapping at a computer keyboard and using its software to write this foreword, I am aware that I am gazing at an electronic machine of great complexity. Yet my very up-to-date computer pales into insignificance when compared with the miracle of natural engineering that is the human body, the living thing that has always fascinated me as a biologist. However much I research or study, there is always something new to discover. Not least of the body's unique features is our large brain which enables me to think up and write these words about the *Encyclopedia of the Human Body*, and for you to read and understand them.

The encyclopedia owes its unique clarity and dynamism to the team that produced it. Expert writers, both doctors and biologists, provided a comprehensive insight into how the body works that is easily understood by children and adults alike. Highly skilled editors and designers presented this information in the best possible way, complementing the text with an incredible array of images that guide the reader on her or his journey through the body. These images include photographs, micrographs, scans, drawings, and diagrams that use colour and detail to, for example, reveal the slimy lining of the intestines, expose the hordes of bacteria on the skin, explain why we do, or do not, have blue eyes, and describe how a baby is born from first contraction to first yell.

To make all this fascinating information accessible for the reader, we have organized the encyclopedia into seven sections. The first five deal with how the body is constructed, how it is supported and moved, how it is controlled, how it maintains itself, and how it reproduces and changes during life. Throughout these sections there are selected pages that focus on specific issues, such as "Understanding the mind" or "Art and anatomy", adding extra breadth to the topics covered in each section.

The final, seventh section contains a detailed timeline and glossary, but it is the sixth section that adds a whole new dimension to this encyclopedia. It explores how understanding of the human body and the development of medicine have gone hand in hand from the earliest times to the modern age. It explains how we know and understand so many of the things we take for granted today – such as how blood circulates round the body or why people get ill – that were a mystery to our ancestors and took all their intelligence and creativity to resolve.

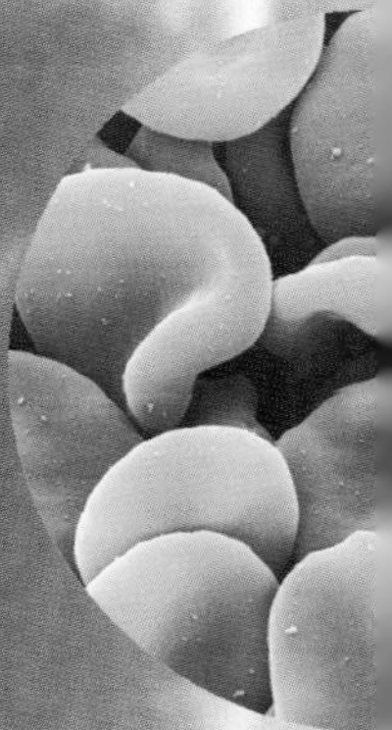

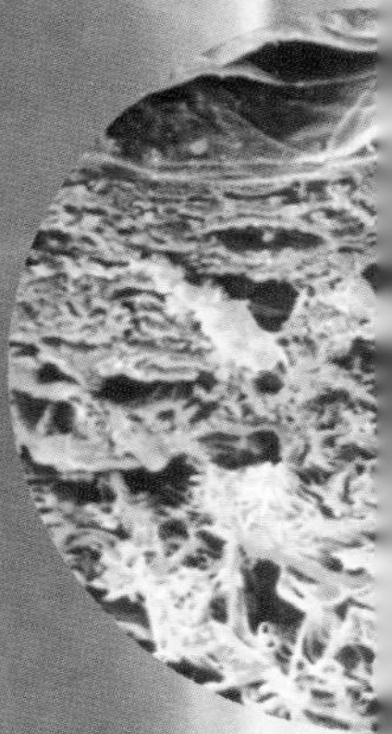

One day, somewhere in the future, a descendant of my desk computer may be smart enough to mimic the thoughts and actions of a human being, but, as you will see when you explore the body's remarkable structure and incredible workings in the *Encyclopedia of the Human Body*, it could never replace the real thing.

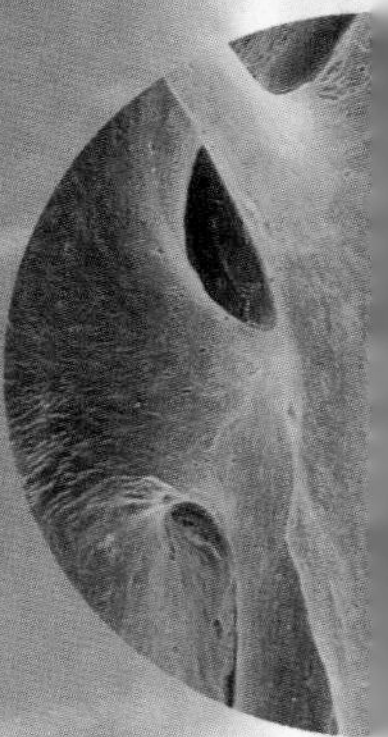

Richard Walker

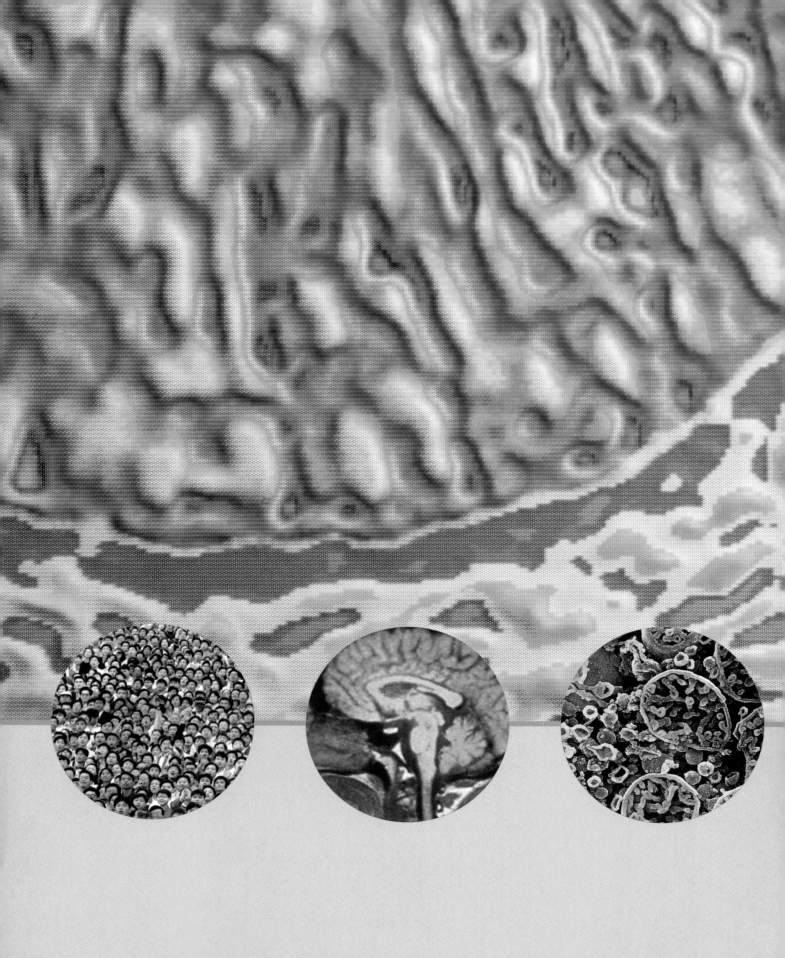

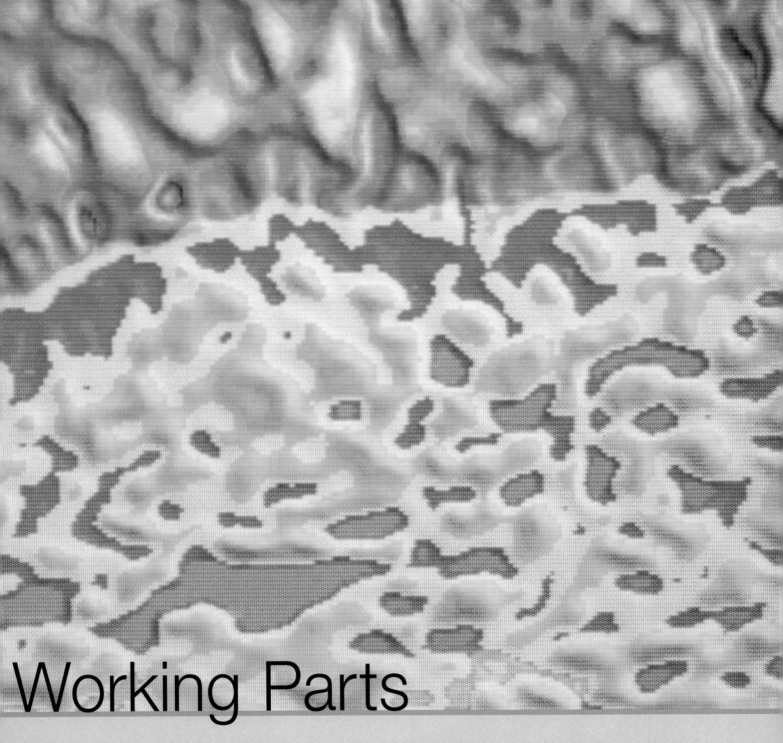

Working Parts

DESPITE THE INCREDIBLE VARIATION in body shapes and sizes, all humans share
the same working parts. At the microscopic level, the human body is built
from huge numbers of cells. Sharing the same basic structure, these tiny
chemical plants are grouped together to make the organs that pump blood,
digest food, take in air, and perform all other life-maintaining functions.
The structure and workings of the living body can be explored, without
cutting it open, using a range of different imaging techniques.

BEING HUMAN

HOMO SAPIENS – Latin for "wise man" – is the name used by scientists to identify you, the reader, and the other six billion humans on planet Earth as a species (type) of animal. It may come as a surprise to some people to realize that humans are actually animals, and just one of the 1.5 million named species that make up the animal kingdom. At the same time, humans stand apart from other animals because their brain power is so superior, even when compared to their close relatives, the apes. Their unique intelligence, communication skills, curiosity, and ability to problem-solve have enabled humans to colonize every continent, triumphing over environments and climates that have defeated other animal species, and to develop a complex understanding of themselves and the world around them.

SOCIAL ANIMALS

Humans are social animals that typically – although not always – live in family units led by a male and female partner, with one or more dependent children. Young humans usually remain with, and are nurtured by, their parents for about 18 years, until they are sufficiently mature and experienced to exist on their own. Several family units together form a larger social grouping or community, which can range in size from a village to a city. In modern industrial and agricultural societies, communities link together to form bigger units, the largest of which – nation states such as the United States or India – contain hundreds of millions of individuals.

Forward-facing eyes *are characteristic of chimpanzees, humans, and all other primates*

Chimpanzee
(*Pan troglodytes*)

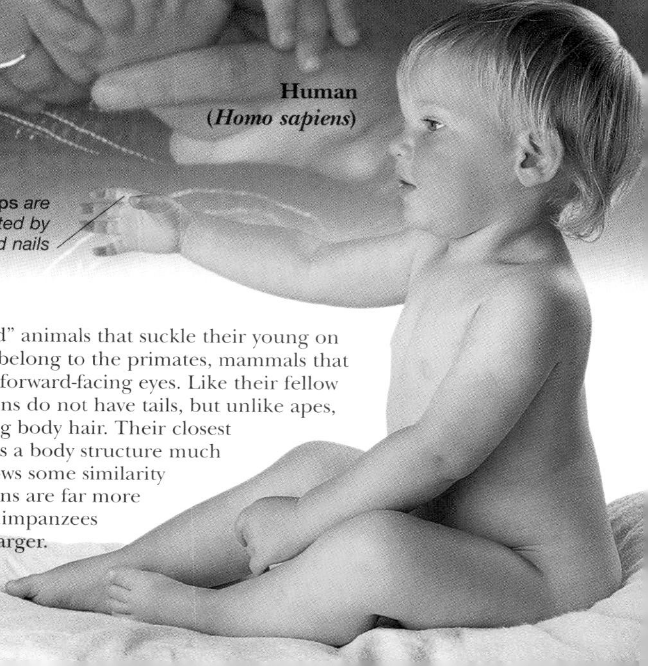

Human
(*Homo sapiens*)

Fingertips *are protected by hard nails*

CLOSEST RELATIVES

Humans are mammals – hairy, "warm-blooded" animals that suckle their young on milk. Along with apes and monkeys, humans belong to the primates, mammals that have five fingers and toes tipped by nails and forward-facing eyes. Like their fellow apes, such as chimpanzees and gorillas, humans do not have tails, but unlike apes, they walk upright and lack long body hair. Their closest relative, the chimpanzee, has a body structure much like that of humans, and shows some similarity in behaviour. However, humans are far more intelligent and skilful than chimpanzees because their brain is much larger.

Inuits wear warm
clothing to survive
the harsh winters

PROTECTIVE CLOTHING

These two Inuit people from Nunavut are wearing thick boots, clothes, and hats to withstand the bitterly cold winters found in northern Canada. Humans are unique among animals in their making and wearing of clothes. This ability allowed humans to leave the confines of tropical Africa, where they first evolved, and move to cooler climates. While clothes have traditionally been made from animal and plant products, such as fur, wool, and cotton, today they are also made from synthetic materials such as nylon. Clothes do more than provide protection from the weather, they also send out messages about the wearer's status in society, their lifestyle, culture, and religious beliefs.

COMMUNICATION

The ability to communicate is essential for humans, as it is for all social animals, to ensure that they live together successfully. Like their ape relatives, humans use body language, gestures, and facial expressions to convey feelings and intentions to each other. But humans also have an additional, unique means of communication – language. By speaking and writing words they can share ideas, plans, decisions, memories, and wisdom. The existence of language means that information can be passed on from one generation to another, providing society with an ever-growing knowledge base.

HUMAN SPECIES FACT FILE

Class	Mammalia (mammals)	Body weight (average)	Males 75 kg (165 lb); females 52 kg (115 lb)
Order	Primates	Activity	Ground-living, diurnal (active during the day)
Species	Homo sapiens		
Distribution	Worldwide	Reproduction	Normally 1 young per litter
Habitat	Most land habitats, living in houses and other shelters	Maximum lifespan	90–100 years
		Conservation status	Not endangered; population increasing worldwide
Food	Animals, plants, and their products		

Keeping in touch by mobile phone

NATURAL VARIATION

IMAGINE STANDING IN a busy city watching the stream of humanity flowing past. One obvious thing an observer would notice is the sheer range of body variation. Everyone, unless they have an identical twin, has a unique combination of features – including their height, weight, shape, hair colour and texture, skin tone, eye colour, and the sound of their voice. However, these are but variations on a central theme that remains constant. Every human body is constructed to a fixed pattern and works in the same way, with minor differences between males and females. Externally, the basic design of a human is an upright body supported by two legs and feet, with two arms and hands that carry and hold, and a flattened face.

BODY REGIONS

Both female and male humans share the same body regions, although their overall shapes and reproductive organs differ. The main axis of the body is made up of the head and trunk. The head contains and protects the brain and sense organs, and is linked by the neck to the trunk, the centre of the body. The thorax, or chest, forms the upper part of the trunk and contains the lungs and heart. The lower part, the abdomen, contains the digestive, reproductive, and urinary organs. Attached to the trunk are the limbs – the arms and legs.

Head contains the brain, which coordinates the body's movements and produces thoughts

Neck holds the head upright and connects it to the trunk

Thoracic cavity, within the thorax, contains the lungs and heart

Abdominal cavity, within the abdomen, contains most digestive organs

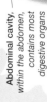

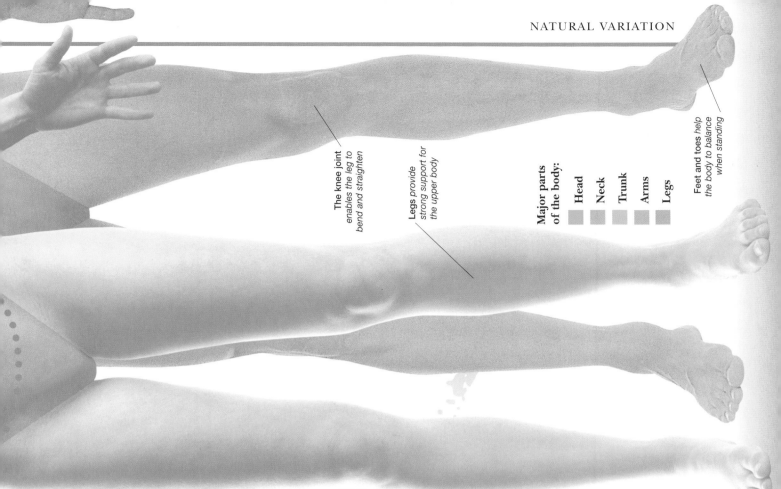

The knee joint
enables the leg to
bend and straighten

Legs provide
strong support for
the upper body

**Major parts
of the body:**

Head

Neck

Trunk

Arms

Legs

Feet and toes help
the body to balance
when standing

Arms are
highly flexible
and can move
in all directions

Hands are free
to manipulate
objects and
use tools

A Spanish
cave painting of
bowmen, dating
from c. 12,000 BC

SEEING OURSELVES

Some 5 million years ago,
human ancestors started to
walk on two legs rather than four.
One consequence of this was that the hands,
no longer involved in supporting the body,
could be used for all kinds of tasks. Over the
millennia, ancient humans gradually became adept at
manipulating objects and using simple tools. As their
brain power increased they developed greater self-awareness,
and humans were able to paint representations of themselves.

HUMAN DIVERSITY

This photo of students in one of the world's most cosmopolitan cities,
London, illustrates clearly the natural variation in human appearance.
Yet the differences in hair colour, skin colour, and facial shape are
relatively minor. These variations have arisen only within the last 70,000
years, as humans spread across the globe from Africa. The differences
reflect ancient adaptations to climatic conditions – for example, dark
skin in the tropical heat, fair skin in cooler areas – that have taken
place during this short time. Today these adaptations have all but
lost their significance as people travel freely between continents.

UNDER THE MICROSCOPE

THE END OF THE 16th century brought with it an invention that would open up a new world for scientists and doctors. In 1590, Dutch instrument maker Zacharias Janssen (1580–1638) made a magnifying device – later to be called the microscope – that, for the first time, made visible objects that were too small to be seen with the naked eye. Initially Janssen's invention made little impact, but in time it would be used to reveal the existence of cells and other previously unseen features of the living world. In the 20th century, the invention of the electron microscope took research into the microworld a step further.

THE DRAPER'S LENS

Following on from Janssen's invention, Dutch cloth merchant Antoni van Leeuwenhoek (1632–1723) made a simple microscope by clamping a tiny convex lens between two brass plates. Although this device sounds primitive, it had a magnifying power of between 70 and 250 times. Van Leeuwenhoek's observations revealed the existence of single-celled organisms (now called protists), as well as some types of body cell. In 1683, he noted minute organisms in his own tooth scrapings, the first bacteria to be seen by the human eye.

SIMPLE MICROSCOPE
Van Leeuwenhoek's microscope (right) was held upright with the eye close to the lens. The object to be examined was placed in front of the lens on a pin which was brought into focus by a series of screws. His drawing shows human sperm, discovered by him in 1677.

DRAWING OF HUMAN SPERM

HOOKE'S DRAWING OF A CORK SECTION

HOOKE'S VIEW
The microscope made by Robert Hooke (left) consisted of a pasteboard barrel, with an eyepiece lens at the top end and an objective lens at the bottom. Here, the specimen is lit by an oil lamp focused through a water-filled sphere. Hooke's sketch of a section through cork shows the chambers he called "cells".

SEM OF LIGHT-SENSITIVE CELLS IN THE RETINA OF THE EYE

COMPOUND MICROSCOPE

Despite van Leeuwenhoek's success, the future of microscopy lay in compound microscopes that used two or more lenses to produce their magnifying effects. British physicist Robert Hooke (1635–1703) made his own compound microscope and, in 1665, presented his observations in a book, illustrated with his own lavish drawings, called *Micrographia*. Among these observations was a section through cork, a dead plant material, showing tiny boxes that he called "cells". This term would later come into common use for a different purpose.

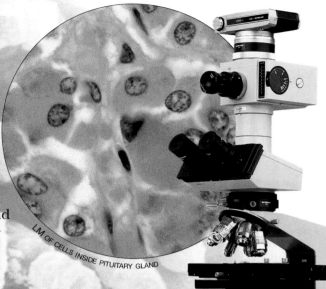

LM OF CELLS INSIDE PITUITARY GLAND

SEEING CELLS

Hooke's microscope and similar instruments of the time were hindered by poor-quality lenses. This defect prevented any great advances being made in microscopy until 1830, when a solution was provided in the form of achromatic lenses, which removed blurring and colour fringes. In the decades that followed, the improved compound microscope was used to discover, identify, and describe human cells and tissues.

THE MODERN VIEW
Used in science laboratories and hospitals worldwide, a modern compound microscope can provide a view of a section through the pituitary gland, as shown here. The otherwise transparent cells have been stained so the viewer can see their component parts.

GREATER MAGNIFICATION

Light, or optical, microscopes have always been restricted by an upper limit of magnification of about 2,000 times. German scientist Ernst Ruska (1906–88) devised an alternative microscope in 1930 that could magnify objects 200,000 times or more. Rather than using light for illumination, it sent a beam of electrons through the thinly sectioned object. This beam – focused not by glass lenses but by electromagnets – was projected onto a phosphor screen, producing a visual image of the object. Called a transmission electron microscope, it revealed among other things the detailed internal structure of cells. The scanning electron microscope, invented in the 1960s, scans the object with an electron beam to produce a 3-D image.

SCANNING ELECTRON MICROSCOPE
A scientist uses a scanning electron microscope to get a 3-D scanning electron micrograph (SEM) of an insect's head.

Cells

LIKE EVERY OTHER living thing on planet Earth, human beings are constructed from tiny, living units called cells. Individually, these cells are transparent, and so minute that they are only visible with a microscope. But the vast numbers of cells present in the body – about 100,000,000,000,000 or 100 trillion – collectively form a recognizable human being. These cells are not all the same; just as within any society certain people play specific roles, so do different types of cell inside the body. This organization is dynamic, not static. Every day the body produces billions of new cells to replace those that have become diseased, damaged, or just worn out. In younger humans, these new cells also enable the body to grow.

Nucleus

Squamous (flat) epithelial cell

LM OF EPITHELIAL CELLS

EPITHELIAL CELLS
The light micrograph above shows epithelial cells from the lining of the cheek. These flat cells fit together like paving stones to protect the inside of the mouth. Other epithelial cells – which may be flat, cube-shaped, or column-shaped – line and protect the body's tubes and cavities, such as blood vessels and the intestines. They also cover the body, forming the epidermis, the upper layer of the skin.

CELL VARIETY
The trillions of cells that make up a person are all derived from a single fertilized egg cell. As this single cell divides repeatedly to make a human being, so groups of cells differentiate. This means that they become specialized, each group taking on its own appearance and role. By the time a person becomes an adult, they will have about 200 different types of cell, each adapted to a particular task. Four of these types – epithelial, bone, sperm, and blood cells – are described here as an example of how body cells differ, and how their appearance is related to their function.

Osteocyte *sits within its own cavity in bone*

Tiny "threads" *link osteocyte to other bone cells*

SEM OF AN OSTEOCYTE

BONE CELLS
Osteocytes are bone cells that live isolated existences within their own tiny cavity, or lacuna, surrounded by the bony matrix that gives the body's bones their strength and hardness. Despite their isolation, these spider-like cells communicate with their neighbours through tiny "threads" that run along minute channels between lacunae. The role of these long-lived cells is day-to-day maintenance of the bony matrix.

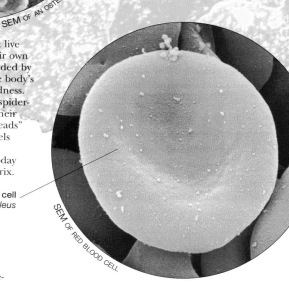

Red blood cell *has no nucleus*

SEM OF RED BLOOD CELL

Tail

Head

SEM OF SPERM

SPERMATOZOA
This scanning electron micrograph shows tadpole-shaped spermatozoa, or sperm, produced by a man's testes. The head of each sperm contains one part of the set of instructions needed to make a new human being; an ovum, or egg cell, produced by a woman, contains the other part. During sexual reproduction, the sperm's tail beats from side to side, pushing the streamlined cell towards an ovum. If the two fuse, a new human life is started.

RED BLOOD CELLS
As their name suggests, red blood cells, or erythrocytes, are found in the blood. Unlike other body cells, these dimpled, doughnut-shaped cells lack a nucleus. Instead they are packed with haemoglobin, a substance that both gives them their colour and carries oxygen. During a four-month lifespan, each red blood cell makes millions of circuits, picking up oxygen in the lungs and delivering it to all parts of the body.

CELL DIVISION

Cells reproduce by cell division. In this micrograph, a cell's nucleus (red) has just divided into two, with the cytoplasm (blue) about to follow. Short-lived cells, such as skin cells, which are constantly worn away, are continually replaced by cell division. At the other extreme, nerve cells do not divide again once they have been formed. The majority of new body cells are produced by a type of cell division called mitosis. Sex cells – sperm and ova – are produced by a process of cell division called meiosis (see pp. 212–13).

Cytoplasm of parent cell divides

CELL THEORY

In 1838, German scientists Jakob Schlieden (1804–81) and Theodor Schwann (1810–82) put forward their Cell Theory, which states that all living things are made of cells. In 1858, German physician Rudolf Virchow (1821–1902) went one stage further by including newly discovered cell division. He stated that new cells can only be made from existing ones, putting paid to the accepted idea that cells could arise spontaneously from non-living material.

Theodor Schwann

One of 23 pairs of chromosomes

Single body cell

New "daughter" cell *is identical to parent*

MITOSIS

This process produces two "daughter" cells that are identical to each other and to their "parent". The control centre of a body cell is called the nucleus. It contains 46 thread-like chromosomes, which hold the genetic information needed to build and run a cell. Before mitosis, each chromosome copies itself. Then, during mitosis, these copies are pulled apart, each to their own new nucleus. The result, once the cytoplasm has divided, is two genetically identical "daughter" cells.

Nucleus of new cell

CANCER CELLS

Normally, cell division is strictly regulated. But if its genetic material is damaged beyond repair, a cell may start to divide uncontrollably and cause a disease called cancer. Once it is out of control, division of a cancer cell (right) produces an expanding clump of abnormal cells called a tumour that can affect normal body functions. Unless the tumour is treated and destroyed, its cancer cells can spread to produce tumours in other parts of the body, eventually causing death.

SEM OF A CANCER CELL

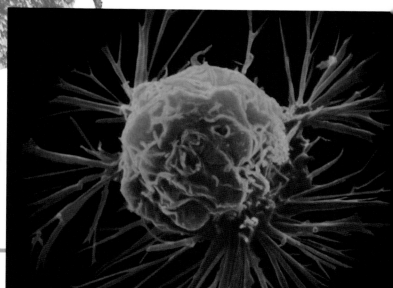

Cell structure

THE FACT THAT CELLS are small does not mean that they are simple. Regardless of their appearance and role, all cells share the same basic, highly organized internal structure. This structure was only revealed with the invention of the electron microscope in the 20th century. Before then, it was assumed that cells were made up of just three components: an outer cell membrane, central nucleus, and – between the two – the cytoplasm. The electron microscope revealed that, rather than being just a featureless jelly, the cytoplasm contained many different components. It is now known that these components work together rather like the different departments in a factory. Some manufacture the materials the cell needs, or recycle substances for reuse. Others generate the energy needed to power the cell's activities. All are controlled by instructions contained within the nucleus.

CYTOPLASM AND NUCLEUS
Lying between the cell membrane on the outside and the nucleus on the inside, cytoplasm consists of a clear, jelly-like fluid – mainly made up of water – called cytosol. Here, organelles, such as mitochondria, float. In the cytoplasm, microfilaments and microtubules form the cell's support system. The nucleus is the cell's control centre, containing the instructions that direct cell activities. The nucleus is surrounded by a nuclear membrane, which has holes, or pores, that allow substances to move between nucleus and cytoplasm.

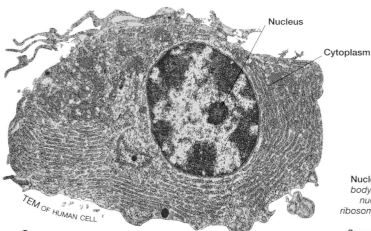

TEM OF HUMAN CELL

Nucleus

Cytoplasm

INSIDE A CELL
Just as each body part, or organ, has its own task, so do the tiny components, or organelles ("little organs"), inside a cell. Largest of these is the nucleus, the cell's control centre. Organelles suspended in the cytoplasm include the endoplasmic reticulum and Golgi body, which manufacture, store, and transport substances, while the mitochondria release energy from glucose and other foods. These cytoplasmic organelles serve to organize the cell's interior into compartments, preventing the cell's many chemical reactions from interfering with each other. Many organelles are surrounded by a membrane, similar in structure to the cell membrane that controls the flow of substances into and out of the cell.

Microfilaments *support and shape the cell*

Nucleolus *is a body within the nucleus where ribosomes are made*

Smooth endoplasmic reticulum *is a system of membrane-enclosed channels where lipids are made*

Secretory vesicle *is a package of substances formed by the Golgi body. It opens at the cell's surface to release its contents*

Nuclear membrane *or envelope forms the boundary of the nucleus*

Cell membrane

Lysosome *contains digestive enzymes that break down substances and worn-out organelles*

Ribosome *is where proteins are made*

Rough endoplasmic reticulum

MITOCHONDRIA

These sausage-shaped organelles (right, in cross-section) are the cell's "power plants", providing the energy needed for many cell activities. Mitochondria (singular form is a mitochondrion) are surrounded by a smooth outer membrane and a folded inner membrane. These folds, called cristae, are where aerobic respiration – the process by which energy is released from food – happens. "Busy" cells – such as liver cells – that need lots of energy contain hundreds of mitochondria.

SEM OF SECTION THROUGH MITOCHONDRIA

ENDOPLASMIC RETICULUM

Extending throughout the cytoplasm is a system of flattened tubules, called endoplasmic reticulum (ER), formed by linked, parallel membranes. ER is responsible for the manufacture, storage, and transport of a range of substances. Rough ER, seen left, is so-called because its surface is studded with granule-like ribosomes, the organelles which make proteins. Newly made proteins are either stored or transported by rough ER.

TEM OF ROUGH ENDOPLASMIC RETICULUM

Peroxisome *contains enzymes that use oxygen to render toxic substances harmless*

Phospholipid molecules

Protein molecule within cell membrane

Mitochondrion

Microtubules *support and shape the cell, and aid movement of substances through the cytoplasm*

Pinocytotic vesicle *through which liquid is taken into the cell*

Golgi body

Model of a "typical" human cell

Cell membrane structure

CELL MEMBRANE

Also called the plasma membrane, this flexible, protective boundary around the outside of the cell, allows some substances to travel in and out of the cell but prevents the movement of others. The cell membrane consists of a "sandwich" of two layers of phospholipid molecules that make it flexible. Proteins dispersed in this double layer have several roles, including transporting food and other substances across the membrane.

GOLGI BODY

Resembling stacked dinner plates, a Golgi body is made up of flattened membrane sacs. It processes proteins produced by rough endoplasmic reticulum and packages them into "bags" called vesicles. These move to the cell membrane and release their contents outside the cell. The Golgi body is most obvious in secretory cells, such as the pancreas cells that release digestive enzymes.

TEM OF GOLGI BODY

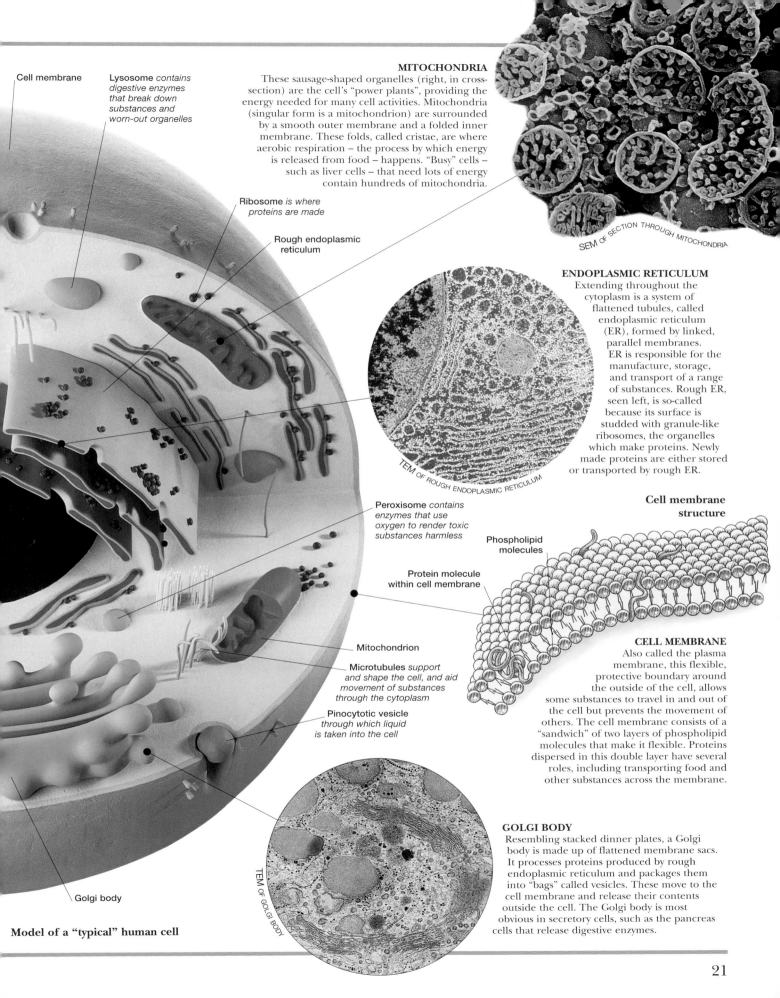

Cell chemistry

INSIDE A PERSON'S CELLS a multitude of chemical reactions is taking place, regardless of whether that person is asleep or awake. During each reaction, chemical compounds are modified to meet the needs of the cell. This mass of chemical activity is far from being chaotic. The organelles of the cell – such as mitochondria – organize chemical reactions into separate compartments so that they do not interfere with each other. Furthermore, reactions are greatly accelerated by special substances called enzymes, which also control how quickly cells consume raw materials and yield new products. Key to a cell's chemistry is the release of energy from food, by the process of cell respiration. Without energy, cells would have no driving force, and life simply could not exist.

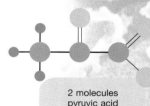

1 molecule
glucose

Glycolysis   ATP

2 molecules
pyruvic acid

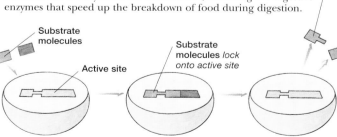

THERMOGRAM OF WOMAN'S MOUTH AND TEETH

CHEMICAL PROCESSES

Inside any cell, two basic processes – catabolism and anabolism – work side-by-side. In catabolism, energy-rich fuel molecules are broken down to release their energy to supply the cell's needs. In fact, only about 25 per cent of this energy can be used by the cell, the rest is released as heat which is used to maintain the body's temperature at 37°C (98.6°F). Anabolism takes simple building blocks to build up the more complex substances a cell needs, such as proteins and lipids. In order to work, anabolism uses the energy generated by catabolism. Together, anabolism and catabolism make up metabolism, the sum of all the chemical processes happening inside a cell.

ENZYMES

These proteins act as catalysts, speeding up the cell's chemical reactions by thousands or millions of times without being changed or used up. Without enzymes, these reactions would take place so slowly that life could not exist. Each enzyme is specific to a particular reaction. Molecules taking part in that reaction bind to the enzyme and react to form product molecules that are then released. Some enzymes work outside cells, including the digestive enzymes that speed up the breakdown of food during digestion.

Product
molecules
move away
from enzyme

Substrate
molecules

Substrate
molecules *lock
onto active site*

Active site

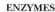

1 *Molecules involved in a reaction, called substrates, fit into a part of an enzyme called the active site.*

2 *The enzyme holds the substrates together. The substrates react and form product molecules.*

3 *The products are released. The enzyme is unaffected by the reaction, and is ready to attract more molecules.*

RAW MATERIALS

Unless it has a constant stream of raw materials to supply its cells' needs, the body cannot survive. During digestion, the carbohydrates, lipids, proteins, and nucleic acids in food are broken down into simple molecules, taken into the blood, and carried to the body's cells. Once inside a cell these are assembled into new molecules by anabolism, or broken down by catabolism. Water is also vital because it provides the liquid medium inside cells in which chemical reactions take place.

AEROBIC RESPIRATION

Body cells get most of their energy from a type of cell respiration called aerobic respiration. Using oxygen, this breaks down fuel molecules such as glucose into carbon dioxide and water to release large amounts of energy. Aerobic respiration has two phases, each made up of several enzyme-catalysed reactions. Firstly, during glycolysis ("glucose splitting") in the cytoplasm, a glucose molecule is split into two molecules of pyruvic acid, with the release of a little energy. Then, inside mitochondria – the cell's "power plants" – a sequence of chemical reactions called the Krebs cycle (see p. 175) completely dismantles pyruvic acid molecules and, with the help of oxygen, releases all of their energy.

6 molecules oxygen

ATP

Krebs cycle

6 molecules carbon dioxide

6 molecules water

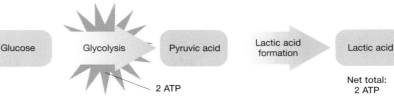

Glucose → Glycolysis → Pyruvic acid → Lactic acid formation → Lactic acid

2 ATP

Net total: 2 ATP

ANAEROBIC RESPIRATION

Cells can temporarily run out of oxygen, for example when muscle fibres (cells) contract rapidly during vigorous exercise. This does not stop cells releasing energy from glucose, however, even though they cannot use aerobic respiration. Instead they use anaerobic ("without air") respiration. In the cytoplasm, glucose is broken down into pyruvic acid – as during glycolysis – which is then converted into lactic acid. This yields much less energy than aerobic respiration, but happens more rapidly. Once oxygen becomes available again, lactic acid is reconverted to pyruvic acid and broken down by the Krebs cycle.

Computer-generated model of ATP

ADENOSINE TRIPHOSPHATE (ATP)

Cells cannot use the energy stored inside glucose until it has been processed by cell respiration. This releases the stored energy in small bursts and uses it to make molecules of adenosine triphosphate (ATP), the cell's main energy store and carrier. When a chemical reaction requires energy, ATP is broken down to liberate its energy store and its components are recycled to pick up more energy from cell respiration. During aerobic respiration, each molecule of glucose yields 38 molecules – "energy packets" – of ATP. The yield from anaerobic is much lower – just 2 ATP.

CHEMICAL COMPONENTS

Human cells – like those of all living things – contain chemical components that are unique to living systems. These are organic compounds, substances whose molecules are based on a "skeleton" of carbon atoms. The main types found in cells are carbohydrates, lipids, proteins, and nucleic acids.

• Carbohydrates provide cells with energy. They include simple sugars, such as glucose, that deliver a source of energy; and complex polysaccharides, such as glycogen, that form fuel stores in the liver and muscle cells.

• Lipids (fats and oils) form the membrane around the cell and its organelles, and provide a long-term energy store in adipose (fat) tissue cells.

• Proteins are complex compounds that perform many different tasks, including forming part of cell membranes, and, as enzymes, controlling cell reactions.

• Nucleic acids such as DNA store in coded form the information needed to control cells by telling them how to make proteins.

Molecular model of glycogen

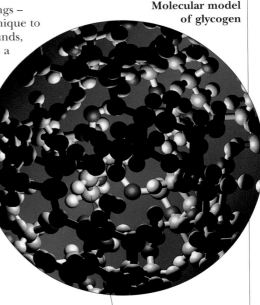

Glycogen *is a complex carbohydrate, or polysaccharide, which is made up of glucose subunits*

THE DOUBLE HELIX

O N 2ND APRIL 1953, the world of science changed forever. Two scientists – American James Watson (b. 1928) and Briton Francis Crick (b. 1916) – announced in the journal *Nature* that they had unravelled the structure of a molecule believed to hold the key to life itself. The molecule was DNA (deoxyribose nucleic acid), which is found in the nucleus of every cell, and its structure they called the "double helix". This new understanding of DNA's structure enabled scientists to find out how it controls the activities of cells and of whole organisms, including humans.

Cell

Chromosome *is made up of tightly coiled DNA*

DNA IN ACTION
Each of the 46 chromosomes in the nucleus of a human cell is made up of a long DNA molecule, sections of which form genes. The two strands of the DNA double helix are linked by bases that always join in a specific way.

Gene *is a section of DNA that carries the instructions for making a specific protein*

"Unzipping" *DNA strands separate so that the bases contained in one strand can be copied*

DNA'S STRUCTURE REVEALED
This photograph shows Watson (left) and Crick next to their model of DNA. Its completion was announced in the journal *Nature* – "We wish to suggest a structure for the salt of deoxyribose nucleic acid (DNA). This structure has novel features which are of considerable biological interest."

Guanine   Thymine

Adenine

mRNA nucleotide

DNA "backbone" *is made of linked phosphate and deoxyribose (sugar) molecules*

Cytosine

Transcription *is where free mRNA bases pair with corresponding bases on DNA to make an mRNA strand*

ROSALIND FRANKLIN
Born in 1920, British scientist Rosalind Franklin played a key role in the discovery of the structure of DNA. Her X-ray diffraction photographs provided Crick and Watson with vital evidence about DNA's spiral shape. Franklin died from cancer in 1958.

DISCOVERING DNA
Early in the 20th century it was realized that cells contain genes, a set of instructions for making and running an organism. Because genes are passed on from parents to offspring, they must be made of a molecule that can both replicate, or copy, itself and hold a store of information. In 1944, American bacteriologist Oswald Avery (1877–1955) showed that it was the nucleic acid DNA which was the carrier of genetic information.

It was known from work carried out during the 1930s that DNA is made up of units called nucleotides. Each nucleotide consists of a phosphate group, the sugar deoxyribose, and one of four nitrogenous bases called adenine, cytosine, guanine, and thymine. In 1953, using evidence from chemical analysis and X-ray diffraction – a method that bounces X-rays off atoms in a DNA of molecule to produce a photographic pattern that indicates its structure – Watson and Crick built their 3-D model of DNA, consisting of two parallel strands that spiral around each other.

UNDERSTANDING THE CODE

Watson and Crick's double helix resembles a twisted ladder. The uprights are made of a "backbone" of phosphate and deoxyribose, the "rungs" of paired bases, adenine with thymine, and cytosine with guanine. They soon realized that this arrangement also provided the means for replication – that the DNA double helix could "unzip", so that free nucleotides could bond to the exposed bases to produce two new DNA double helices. But how does DNA control cell activities?

Since the 1940s it had been known that genes control the production of proteins, many of which are enzymes, the biological catalysts that regulate chemical reactions inside cells. Each protein is made up of a specific sequence of amino acids,

COPYING THE MESSAGE

DNA has the unique ability to make an exact copy of itself. This TEM shows a DNA strand (shown right in yellow) that is "unzipping" to form two single "daughter" strands, each of which acts as a template to form a new double helix identical to the "parent".

DNA strand *has "unzipped" into two strands*

TEM OF REPLICATING DNA STRAND

MAKING PROTEINS

Protein synthesis happens in the cytoplasm of the cell, but DNA molecules are too large to move out of the nucleus. So how is the message conveyed from nucleus to cytoplasm? A section of DNA (a gene) "unzips", and one strand is copied (transcription) by a smaller, single-stranded nucleic acid called messenger RNA (mRNA). This contains the same bases as DNA, except for uracil, which replaces thymine. The mRNA passes into the cytoplasm and attaches itself to a ribosome. Here, the mRNA is translated triplet by triplet so that amino acids are linked up in the correct sequence to make a specific protein.

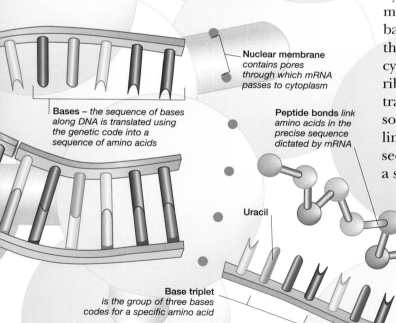

Nuclear membrane *contains pores through which mRNA passes to cytoplasm*

Bases – *the sequence of bases along DNA is translated using the genetic code into a sequence of amino acids*

Peptide bonds *link amino acids in the precise sequence dictated by mRNA*

Uracil

Base triplet *is the group of three bases codes for a specific amino acid*

Free amino acids *in the cytoplasm are "assembled" to make a particular protein*

Newly assembled protein *detaches itself from the ribosome and folds up, its functional shape determined by the specific sequence of amino acids*

mRNA

Ribosome *provides a site on which the mRNA message is translated into protein structure*

of which there are 20 different types. During the 1960s Marshall Nirenberg (b. 1927), an American biochemist, showed that within DNA an arrangement of three bases specifies one particular amino acid. This provides a genetic code in which base triplets form the "words" that instruct the cell to select the right amino acids to make a specific protein.

TRANSLATION

An organelle called a ribosome passes along the strand of mRNA using the genetic code to translate it into protein structure. Amino acids are lined up in the correct sequence, and then bond together to form a protein.

Tissues

IF THE BODY'S TRILLIONS of cells all existed independently, it would be impossible to organize and operate a living human being. Instead, cells of the same or similar types are grouped together into tight-knit communities in which they work together to carry out a specific task. These groups of similar cells form tissues – a word derived from the Latin for "woven" – just as cotton threads are interwoven to make cloth. The many different types of tissue fall into four basic categories – epithelial, connective, muscular, and nervous – that interface to make up the fabric of the body. If tissues are damaged, their cells divide to repair the damage. This process, called regeneration, happens more quickly in some tissues than others. Together, different types of tissue form organs such as the heart and kidneys, as described in more detail on pp. 28–29. The roles of the four basic types of tissue can be described very simply as follows: epithelial tissues cover; connective tissues support; muscle tissues move; and nervous tissues control.

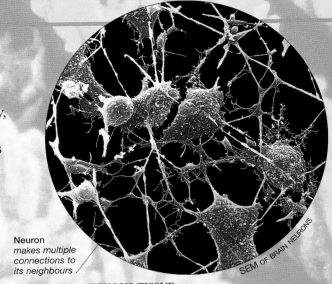

SEM OF BRAIN NEURONS

Neuron *makes multiple connections to its neighbours*

NERVOUS TISSUE

Restricted to the nervous system – the brain, spinal cord, and nerves – nervous tissue is responsible for controlling and coordinating most body processes, as well as providing humans with conscious thought and sensation. Two types of cell are found in nervous tissue. Neurons, or nerve cells, like the brain neurons above, enable communication to take place by generating and carrying electrical signals at high speed. Glial cells, or neuroglia, support and nurture the neurons.

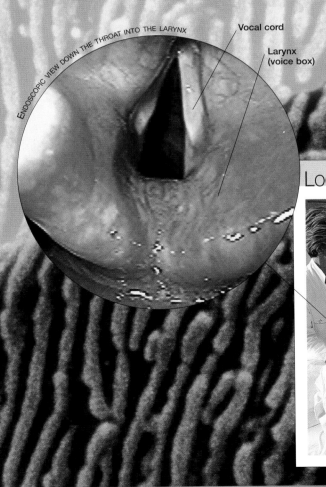

ENDOSCOPIC VIEW DOWN THE THROAT INTO THE LARYNX

Vocal cord

Larynx (voice box)

LOOKING INSIDE

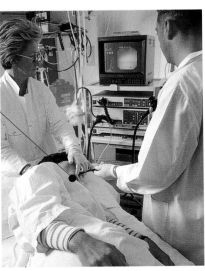

Doctors can look inside the body to examine tissues and organs by using a technique called endoscopy. Endoscopes are tube-like instruments that contain long optical fibres which transmit light. The endoscope can be inserted through a natural opening, such as the mouth, as shown here, or through a small incision made in the skin. Optical fibres carry light to illuminate the tissue being observed, while a miniature camera relays images to a screen for the doctor to see. The endoscope may also include tiny scissors or forceps to take a small sample of tissue for biopsy. Tissue is then looked at under the microscope for signs of disease.

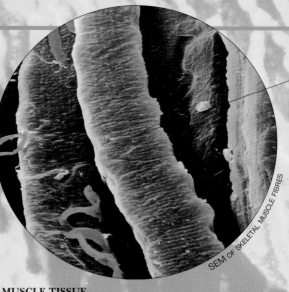

Muscle fibre

SEM OF SKELETAL MUSCLE FIBRES

Chondrocyte
*is a mature
cartilage cell*

CONNECTIVE TISSUES

Unlike other types of tissue, connective tissue consists of cells embedded in a matrix, or framework, which is secreted by the cells of the tissue. With major roles including binding and support, protection and insulation, connective tissues are the most abundant and diverse of all body tissues. They hold the body together by underlying epithelial tissues and packaging organs, and are also found in tendons, ligaments, and the dermis of the skin. Cartilage (left, seen inside a joint) and bone in the skeleton support and protect the body's organs. Adipose tissue, its cells packed with fat, insulates and cushions organs. Blood, a connective tissue with a liquid matrix, transports materials around the body.

Cartilage matrix *is
rich in collagen fibres*

LM OF SECTION THROUGH HYALINE CARTILAGE

MUSCLE TISSUE

The body's muscular tissues consist of cells, called fibres, that can contract, or shorten. Skeletal muscle tissue, seen in this micrograph, moves the body and maintains its posture. Smooth muscle tissue, in the walls of hollow organs, typically moves materials through the body. Cardiac muscle tissue, found in the wall of the heart, pumps blood around the body. Muscular tissues receive a rich blood supply which brings the food and oxygen needed to release energy for contraction.

EPITHELIAL TISSUE

Also called epithelium, epithelial tissue consists of a continuous sheet of cells, which may be one cell or many layers thick. It forms the outer layer of skin, and the inner linings of the digestive system – including the folded lining of the oesophagus, shown here – and of the respiratory, urinary, and reproductive systems, blood vessels and the heart. By covering and lining, epithelial tissues protect the body's surfaces, form a barrier to micro-organisms, and provide an interface through which all substances entering or leaving the body must pass.

SEM OF LINING OF OESOPHAGUS

LM OF SECTION THROUGH NASAL MUCOUS MEMBRANE

MEMBRANES

Together, epithelial and connective tissues combine to form the membranes that line hollow organs and body cavities. The micrograph (above) shows a section through the membrane that lines the nasal cavity. Epithelial tissue (blue-pink) secretes sticky mucus that traps dust particles in the air. The connective tissue (green-gold) is reinforced by tough collagen fibres and stretchy elastin fibres, which underpin and stabilize the epithelial tissue above it.

Organs and systems

LOOKING AT THE pictures of living bodies on these pages, it is as if a door to each body's interior has been swung open to expose the brain, liver, and other organs. But, however revealing these images may be, they provide no insight into how different organs work together to produce a living human. In fact, just as individual players in a soccer team collaborate to score goals, specific body organs operate as a team – called a system – to carry out a specific task. Digestive system organs, for example, interact to supply the body with food. In all, there are 12 systems, but they do not exist in isolation. Instead they work together to make a complete, walking, talking human being.

Cellular level
These are four types of cells found in the epithelial tissue that lines the stomach, and which together contribute to the digestive process.

Tissue level
*This section through the stomach wall shows three types of tissues:
(1) epithelial,
(2) connective,
(3) muscular.
Each is made up of specific types of cells.*

Larynx (voice box) *is the part of the respiratory system that produces sounds*

Kidney *removes wastes and excess water from the blood to form urine*

Organ level
Like other organs, the stomach has a recognizable shape and is made up of different tissues that work together so it can perform specific tasks.

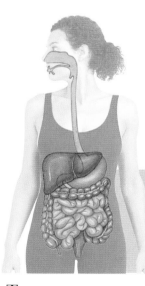

System level
The stomach, small intestine, and other digestive organs are linked together to form the digestive system, which is responsible for getting food into the body. Together, this and other systems make up the body.

THE BODY HIERARCHY

The organization of the body can be seen as a sequence of levels, starting with the simplest – the body's atoms and molecules – and progressing to the most complex, the body itself. Key molecules – such as proteins and lipids – make up cells and their organelles. Similar cells that work together make up a tissue, while two or more tissues make up an organ. The stomach, for example, has a lining of epithelial tissue, layers of muscle tissue which contract to crush food during digestion, and connective tissue which holds the organ together. Linked organs work together to form a system, which interacts with the other systems to make up the body.

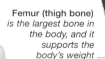

Femur (thigh bone) *is the largest bone in the body, and it supports the body's weight*

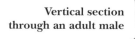

Vertical section through an adult male

WHAT IS AN ORGAN?

A glance into a mirror immediately reveals some organs. Eyes and skin are the most obvious, while inside an open mouth are the tongue, teeth, and tonsils. However, most organs are found inside the body, as these magnetic resonance imaging (MRI) scans show. Here the brain, lungs, kidneys, and liver can be seen in a vertical section, as can individual bones and muscles. Each organ is made of two or more tissues (a group of cells of the same type) and carries out one or more specific tasks that are essential for the body's survival.

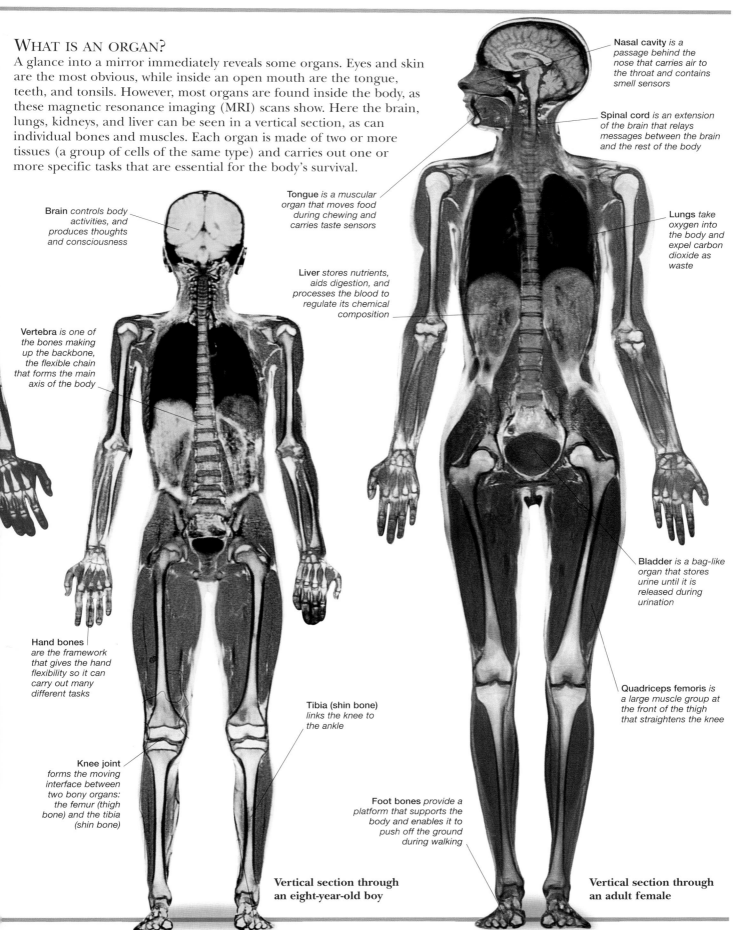

Nasal cavity *is a passage behind the nose that carries air to the throat and contains smell sensors*

Spinal cord *is an extension of the brain that relays messages between the brain and the rest of the body*

Brain *controls body activities, and produces thoughts and consciousness*

Tongue *is a muscular organ that moves food during chewing and carries taste sensors*

Lungs *take oxygen into the body and expel carbon dioxide as waste*

Liver *stores nutrients, aids digestion, and processes the blood to regulate its chemical composition*

Vertebra *is one of the bones making up the backbone, the flexible chain that forms the main axis of the body*

Hand bones *are the framework that gives the hand flexibility so it can carry out many different tasks*

Knee joint *forms the moving interface between two bony organs: the femur (thigh bone) and the tibia (shin bone)*

Tibia (shin bone) *links the knee to the ankle*

Bladder *is a bag-like organ that stores urine until it is released during urination*

Quadriceps femoris *is a large muscle group at the front of the thigh that straightens the knee*

Foot bones *provide a platform that supports the body and enables it to push off the ground during walking*

Vertical section through an eight-year-old boy

Vertical section through an adult female

Imaging techniques

TODAY, DOCTORS AND researchers have access to many different methods of looking inside a living body. This allows them to explore tissues and organs to search for disease, or to find out how the body works. This has not always been the case. Until 40 years ago, X-rays – which do not clearly reveal softer body tissues – were about the only means of seeing inside a living person without having to perform surgery. Since the 1970s, however, modern technology, especially advances in computers, has produced a variety of powerful imaging techniques, including CT and MRI scanning. Like X-rays, these techniques are non-invasive (do not require surgery), but they produce images – many examples of which feature throughout this encyclopedia – that are considerably more detailed than anything obtained before.

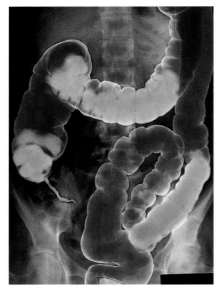

X-RAYS (RADIOGRAPHY)
With this technique, a high-energy form of radiation is passed through the body and projected onto a photographic film to produce an X-ray photograph, or radiograph. The film remains white where hard tissues, such as bone, have absorbed X-rays, but turns grey or black where X-rays have passed through soft tissues, such as muscle. In this contrast radiograph, barium sulphate – a substance that absorbs X-rays – has been introduced into the large intestine (made of soft tissues) so its outline (orange) can be clearly seen. The colours are false and were added afterwards.

COMPUTED TOMOGRAPHY (CT)
Combining X-rays with a computer, CT scanning produces much more detailed images than ordinary X-ray photographs. As a person lies inside a scanner (below), it rotates around him sending narrow beams of X-rays through his body and into a detector. A computer analyzes information from the detector to produce a "slice" through the organs in that part of the body. These slices can be built up to produce a 3-D image, like this one (right) of the skull and brain.

DIFFERENT METHODS

Imaging techniques vary in the way they work and in their applications. X-rays and CT scanning use high-energy radiation. Ordinary X-rays are typically used for looking at bones. Contrast X-rays use special substances to reveal hollow structures such as the intestines or blood vessels. CT scanning uses computers to produce detailed images, usually of the head or abdomen. PET and radionuclide scanning both use radioactive substances to reveal chemical activity in tissues rather than detailed structure. MRI scanning and ultrasound do not use radiation. MRI scans produce detailed images of any body tissue. Ultrasound can show both structure and movement, and is commonly used to observe the development of the fetus.

Patient undergoing CT scan

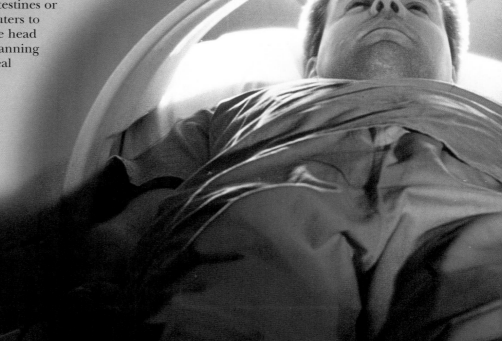

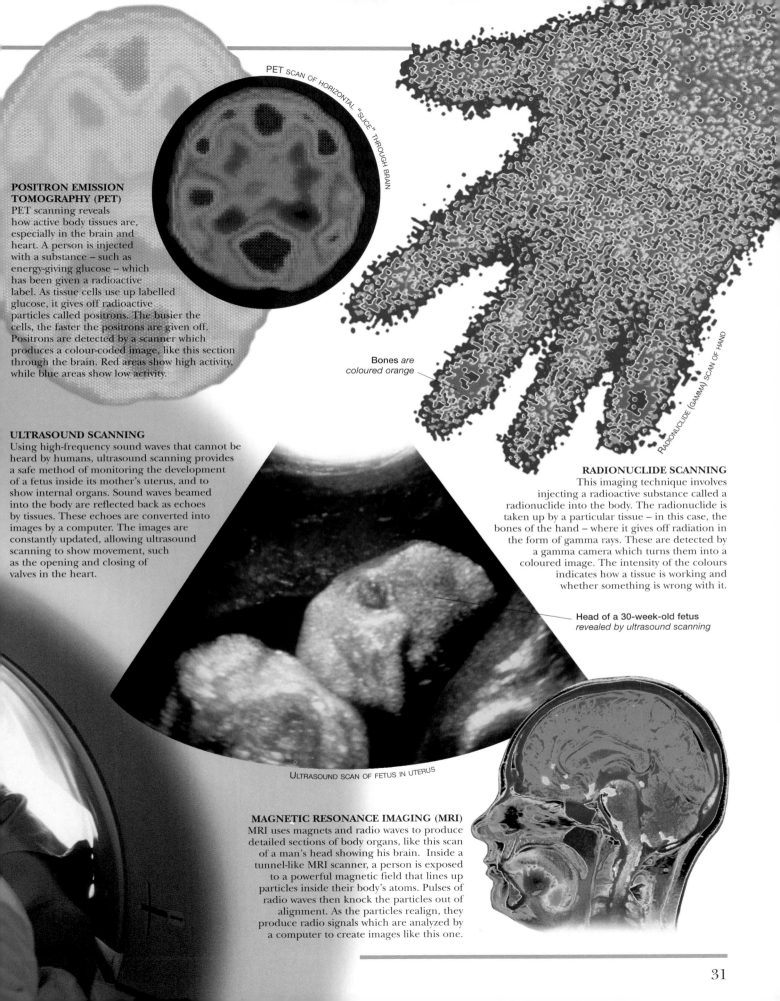

POSITRON EMISSION TOMOGRAPHY (PET)

PET scanning reveals how active body tissues are, especially in the brain and heart. A person is injected with a substance – such as energy-giving glucose – which has been given a radioactive label. As tissue cells use up labelled glucose, it gives off radioactive particles called positrons. The busier the cells, the faster the positrons are given off. Positrons are detected by a scanner which produces a colour-coded image, like this section through the brain. Red areas show high activity, while blue areas show low activity.

ULTRASOUND SCANNING

Using high-frequency sound waves that cannot be heard by humans, ultrasound scanning provides a safe method of monitoring the development of a fetus inside its mother's uterus, and to show internal organs. Sound waves beamed into the body are reflected back as echoes by tissues. These echoes are converted into images by a computer. The images are constantly updated, allowing ultrasound scanning to show movement, such as the opening and closing of valves in the heart.

Bones *are coloured orange*

RADIONUCLIDE SCANNING

This imaging technique involves injecting a radioactive substance called a radionuclide into the body. The radionuclide is taken up by a particular tissue – in this case, the bones of the hand – where it gives off radiation in the form of gamma rays. These are detected by a gamma camera which turns them into a coloured image. The intensity of the colours indicates how a tissue is working and whether something is wrong with it.

Head of a 30-week-old fetus *revealed by ultrasound scanning*

ULTRASOUND SCAN OF FETUS IN UTERUS

MAGNETIC RESONANCE IMAGING (MRI)

MRI uses magnets and radio waves to produce detailed sections of body organs, like this scan of a man's head showing his brain. Inside a tunnel-like MRI scanner, a person is exposed to a powerful magnetic field that lines up particles inside their body's atoms. Pulses of radio waves then knock the particles out of alignment. As the particles realign, they produce radio signals which are analyzed by a computer to create images like this one.

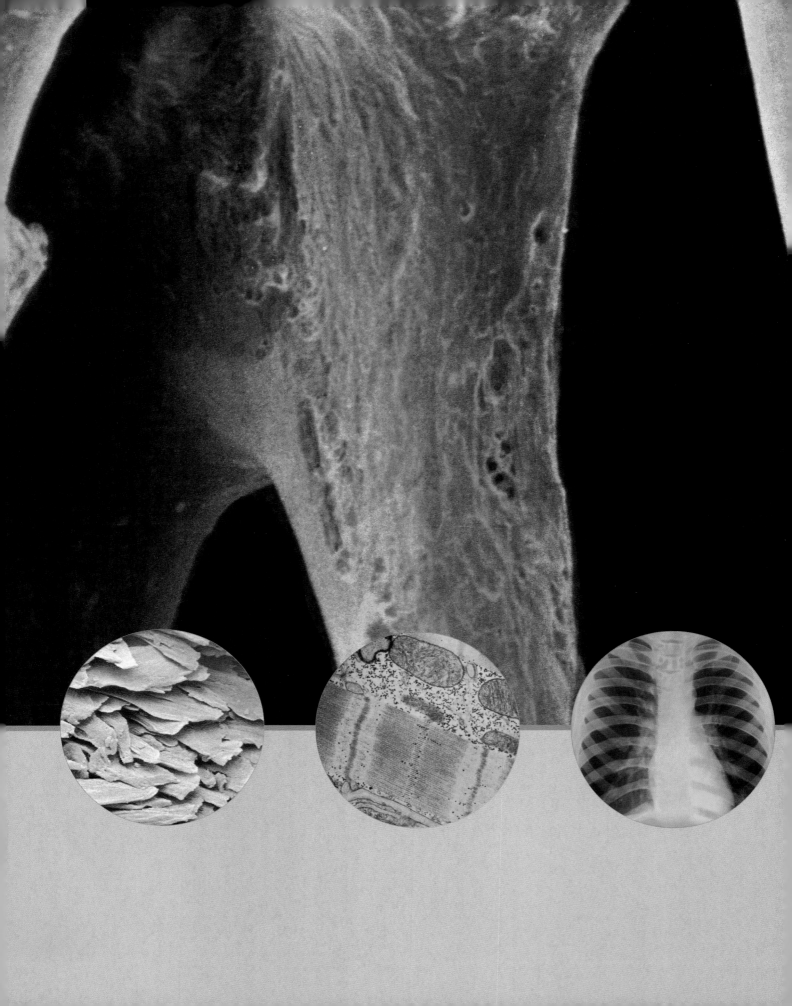

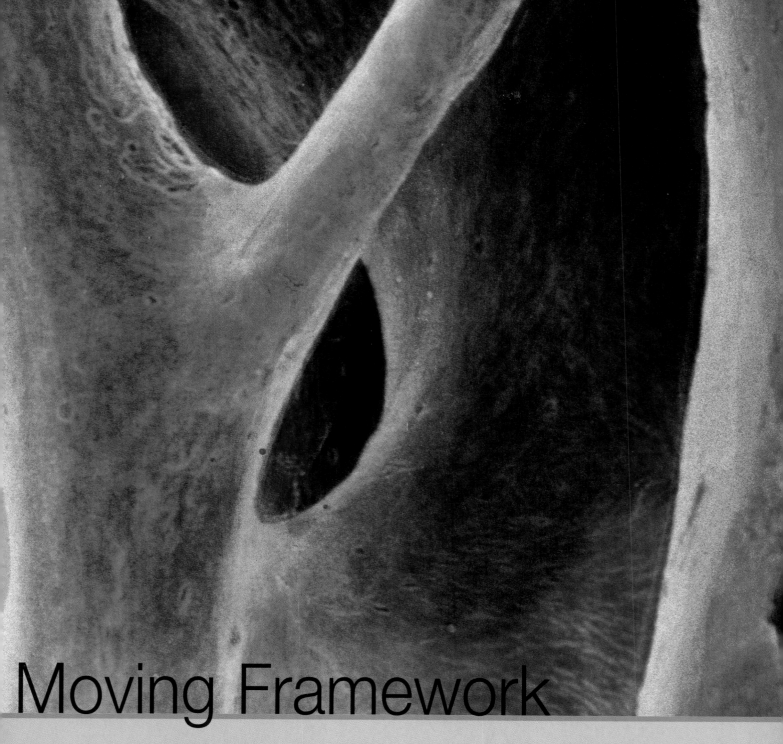

Moving Framework

THREE SYSTEMS COVER, move, shape, and support the body. Skin is a living, stretchy overcoat that shields the delicate tissues inside the body from the harsh conditions outside. Bones form the skeleton, a structure strong enough to support the body's weight and prevent its collapse, yet light and flexible enough to let the body move. Muscles shape the body, and, by pulling bones, produce a multitude of movements from raising an eyebrow to running a marathon.

INTEGUMENTARY system

Cornified layer *consists of flat, dead cells that are constantly worn away*

Clear layer *is most apparent in the thick skin that covers the soles and palms*

Granular layer *is where cells flatten and fill with tough keratin as they move towards the surface cells*

Basal layer *produces new cells to replace those lost from the surface*

Spiny layer *cells are linked by spine-like connections*

Dermal papillae *attach dermis to the epidermis*

SKIN, HAIR, AND NAILS make up the integumentary system. The skin, which covers the entire surface of the body and forms a vital barrier between the body and its surroundings, is the body's largest organ. In an average adult, it weighs about 5 kg (11 lb). It provides protection against injury, infection by micro-organisms, and damage by harmful rays in sunlight. The skin is also a sense organ that can detect touch, warmth, cold, and pain. In addition, it helps to control body temperature and produces vitamin D, which is necessary for healthy bones. Hair and nails grow directly from the skin to provide additional covering and protection.

LIVING LAYERS

The skin has two layers. The upper layer, the epidermis, is made up of cell layers (left, shown separated) that become flatter and tougher towards the surface. The lower layer, the dermis, contains strong, flexible fibres, blood vessels, nerves, and sensory receptors. Lodged in the dermis are sebaceous glands and coiled sweat glands. Hairs grow from follicles that extend upwards from the dermis through the epidermis. Beneath the dermis lies subcutaneous tissue, a fatty layer that insulates the body and stores energy.

Nerve *relays messages between skin receptors and the brain*

Arrector pili muscle *pulls hair upright to produce goose bumps*

Sebaceous gland *produces oily sebum which keeps skin and hair soft and flexible*

Subcutaneous fat *helps to insulate the body*

Touch receptor *detects light touch*

Sweat gland *releases sweat onto the skin's surface to cool the body*

Pressure receptor *detects pressure and vibrations*

Blood vessel *helps to regulate body temperature*

Hair follicle *is a cavity in the skin from which hair grows*

Section through the skin

PROTECTIVE OVERCOAT

The epidermis provides a waterproof coating that protects the body from drying out or becoming waterlogged. Water is repelled from the skin's surface by the keratin that fills epidermal cells and by oily sebum produced by the sebaceous glands. As a person showers, skin keeps the water out. Meanwhile, receptors that are sensitive to heat or cold detect the temperature of the water, and receptors that are sensitive to touch detect the flow of water against the skin.

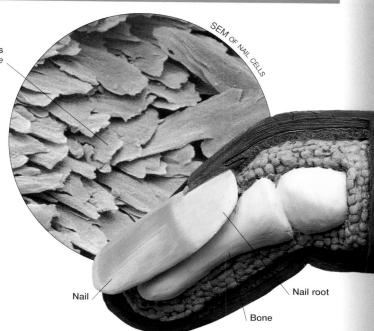

SEM OF NAIL CELLS

Flattened dead cells
from nail surface

Nail

Nail root

Bone

Section through a finger showing position of nail

INTEGUMENTARY SYSTEM FUNCTIONS

Protection	*Against the harmful effects of physical injury, chemicals, heat, sunlight, infection, and excessive water; also against water loss.*	Vitamin D synthesis	*Produces vitamin D in the presence of ultraviolet rays from the sun.*
		Excretion	*Eliminates small amounts of waste substances from the body in sweat.*
Temperature regulation	*Keeps body temperature stable by sweating and varying size of blood vessels in the skin; vessels narrow to conserve heat and widen to lose heat; hair limits heat loss from the head.*	Gripping	*Provides a surface with good grip for handling objects and to prevent slipping; nails make handling easier.*
		Absorption	*Can take small amounts of certain substances into the body from the surface.*
Sensation	*Detects touch/pressure, pain, warmth, and cold.*		

NAILS

The sensitive tips of the fingers and toes have hard plates called nails that provide protection and support. Each nail grows from a living root at its base, which is overlapped by a fold of skin, and consists mainly of flattened dead cells filled with the tough protein keratin. Nails grow faster in warm weather than cold, and fingernails grow three or four times faster than toenails.

"NAKED" APES

Humans appear naked compared to great apes, our closest animal relatives. However, apes and humans have similar numbers of body hairs. Whereas the ape's body is coated with long, coarse hair (terminal hair), the human body is mostly covered with short, fine hair (vellus). Humans have terminal hair on the scalp and in a few other areas only. A hairy coat helps keep the ape warm; humans rely mainly on clothing instead.

Light, downy vellus hair

Coarse terminal hair

Skin surface

COVERING AN AREA of up to 2 sq m (21.5 sq ft), the skin's surface is far from smooth. In fact, it is marked by numerous criss-crossing lines and by creases, grooves, ridges, and bumps. It is kept soft and supple by a thin coating of oil, called sebum. Scattered over the surface are the openings of millions of hair follicles and sweat ducts. Sweating is one of the ways in which the skin helps to control body temperature. The surface of the skin is also home to a variety of bacteria, visible only under a microscope. Sometimes large numbers flourish, causing spots or rashes.

Goose bumps

COOLING DOWN

If the body overheats, droplets of sweat ooze from sweat glands on to the skin's surface. Water in the sweat evaporates and draws heat away from the body, cooling it down. The sweat glands in the armpits produce a large amount of sweat, which many adults use antiperspirants to control. Cooling is also achieved by blood vessels in the skin, which widen to speed up heat loss from the body.

SEM OF SWEAT DROPLETS ON SKIN

GOOSE BUMPS

When the body is cold, blood vessels in the skin narrow to conserve heat and small bumps, called goose bumps, appear. Goose bumps are formed when a tiny muscle attached to the base of each hair shortens, pulling the hairs upright and lifting the surrounding skin. In animals with thick coats, air becomes trapped between the upright hairs, creating a blanket that helps keep the body warm.

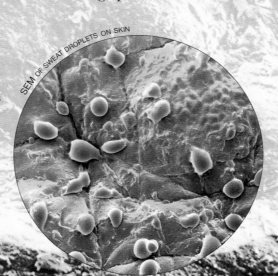

EPIDERMAL RIDGES

This close-up view shows the hundreds of tiny ridges on the surface of the skin on the palm of a hand. Sweat ducts open in rows along the crests of each ridge. The undersides of hands and feet are the only areas covered with these ridges, which are separated by fine parallel grooves and form curved patterns on the skin. They are also the only areas that have no hair or oil glands. A ridged, hairless surface provides good grip for handling objects and prevents slipping on surfaces, such as when walking or climbing.

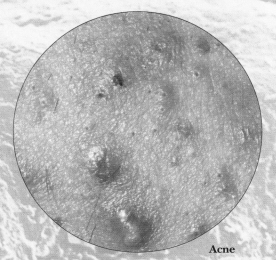

Acne

SEM OF BACTERIA ON SKIN

SPOTS AND PIMPLES
Bacteria or other micro-organisms can infect the skin, resulting in spots and pimples. Skin problems are common during adolescence, when changes in hormone levels cause sebaceous glands to produce excess sebum which becomes trapped in hair follicles. Skin bacteria thrive in the sebum, causing inflammation of surrounding tissues, and often giving rise to acne, as shown above on the face of a teenage boy.

SKIN BACTERIA
This magnified view shows some of the millions of harmless bacteria that normally live on the skin. These bacteria help to keep the skin healthy by preventing harmful varieties from growing. The oil on the surface of the skin also helps eliminate harmful bacteria. Fungi are often present in deep folds in the skin, such as between the toes or in the groin.

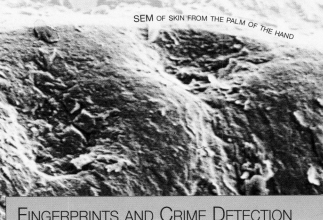

SEM OF SKIN FROM THE PALM OF THE HAND

FINGERPRINTS AND CRIME DETECTION

The swirling epidermal ridges at the ends of the fingers and thumbs produce patterns that are unique to each individual. Well supplied with sweat glands, these ridges leave behind sweat patterns, better known as fingerprints, when they touch smooth surfaces. The unique nature of fingerprints makes them useful in crime detection. Investigators look for similarities between fingerprints found at a crime scene and those of suspects. During the 1880s, interest in the use of fingerprints grew, and London's Metropolitan Police set up a fingerprint department in 1901 to help identify criminals. In 1902, a burglar named Jackson became the first criminal to be convicted on the basis of fingerprint evidence. In the United States, the first fingerprint file was established by J. Edgar Hoover, who became director of the FBI in 1924.

A fingerprint, with its unique pattern of whorls and arches

Skin features

TOGETHER, THE TWO layers of the skin – the epidermis and dermis – combine to give it depth. This can vary considerably – the skin is very thin and delicate in some areas, and much thicker and tougher in others. Regardless of its thickness, skin everywhere undergoes a continuous process of renewal. Dead, flattened cells are constantly worn away to be replaced by new cells generated by the division of living cells in the lowest layer of the epidermis. Scattered within this multiplying layer are cells that produce melanin, the brown pigment that helps to give skin its colour.

BRUISES
Bruising – which is usually caused by a knock or fall – occurs when blood leaks from damaged blood vessels. It can be very noticeable around the eye because here the skin is not only very thin but also loosely anchored, so blood can easily collect – producing a black eye. Bruises usually look dark purple or blue at first, then turn brown, green, or yellow as they fade.

TWO LAYERS

The SEM below shows the two skin layers: the epidermis above, and the thicker dermis beneath. The epidermis contains flat, overlapping cells that are filled with a tough protein called keratin and which form a protective, waterproof covering. Below, the dermis contains collagen and elastin fibres which give the skin strength and flexibility, allowing it to stretch and return to its normal shape. The dermis is also supplied with nerves and blood vessels.

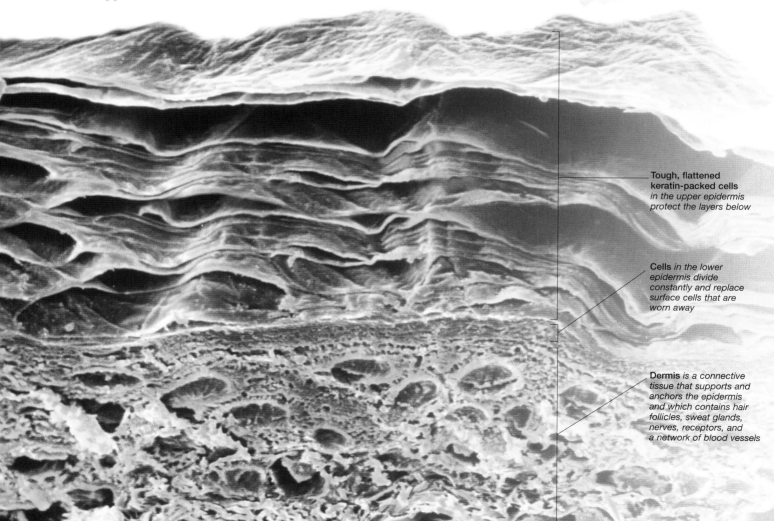

Tough, flattened keratin-packed cells *in the upper epidermis protect the layers below*

Cells *in the lower epidermis divide constantly and replace surface cells that are worn away*

Dermis *is a connective tissue that supports and anchors the epidermis and which contains hair follicles, sweat glands, nerves, receptors, and a network of blood vessels*

Sole of the foot: 4 mm (0.16 in)

Eyelid: 0.5 mm (0.02 in)

HOW THICK?

Skin varies in thickness over the surface of the body. It is very thin on the eyelids and lips, and very thick on the soles of the feet and palms of the hands. The epidermis of the skin tends to become thicker and harder if it experiences a lot of wear and tear. For example, people who often walk barefoot have tougher soles of the feet.

SUNTAN

Sunlight stimulates the skin to make more melanin – which is why skin darkens, or tans, in the sun. The extra melanin provides increased protection against harmful ultraviolet rays from the sun. Too much ultraviolet radiation can cause sunburn. It can also make the skin appear dry and wrinkled, and increases the risk of developing skin cancer in later life. Wearing a hat, and applying sunscreens regularly, helps to reduce the sun's effects on the skin.

An Australian lifeguard wearing a hat and sunscreen

SKIN FLAKES

Tens of thousands of dead, flattened cells are shed from the skin every minute. They constantly rub off or peel away from the skin surface, like flakes of old paint, to be replaced by new cells that push up from the lower epidermis towards the surface. These fallen flakes of dead skin, together with other particles and fibres, form household dust.

SEM OF SKIN FLAKES

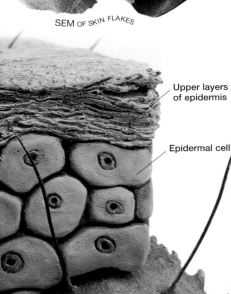

Upper layers of epidermis

Epidermal cell

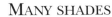

Pregnant female scabies mite *burrows into the epidermis*

MANY SHADES

Skin colour ranges from nearly black through varying shades of brown to pale pink, and depends on how much melanin the skin contains. Melanin is a dark pigment made by cells called melanocytes in the lower epidermis, and serves to protect the skin against damage by ultraviolet rays. The amount each person has is determined by their ancestry. People with dark skin have a lot of melanin, people with fair skin have less.

ITCH MITES

The itch mite has been a human parasite for thousands of years. The tiny female mite burrows deep into the epidermis, where she lays her eggs, and a few weeks later an intensely itchy rash – called scabies – develops on the trunk and limbs. Passed on via close contact, such as holding hands with an infected person, the scabies mite can only be killed by special medicated lotions and creams.

Hair

ALMOST EVERY PART OF THE body – apart from the palms of the hands, soles of the feet, lips, and nipples – is covered by hair. Hairs are long filament-like structures made up of dead cells that grow out of the skin. Short, fine vellus hair covers much of the body, while longer, thicker terminal hair is found in the eyebrows, eyelashes, nose hairs, and on the scalp. After puberty, terminal hair appears in the armpits and pubic regions of both sexes, as well as on the faces and chests of males. There are about 100,000 hairs on the scalp, of which about 100 are lost daily to be replaced by new growth. Scalp hair serves to insulate the head and protects it from harmful sunlight radiation.

HEAD OF HAIR

Whether hair is straight, curly, or wavy depends on the shape of its shaft. In cross- section, shafts of straight hair are round, those of wavy hair are oval, and those of curly hair flat. Hair colour depends on how much of each of the three melanin variants – yellow, red, and brown-black – hair contains. The relative amounts of each determines whether hair is blond, red, brown, or black. In older people, a slow-down in melanin production produces grey hair.

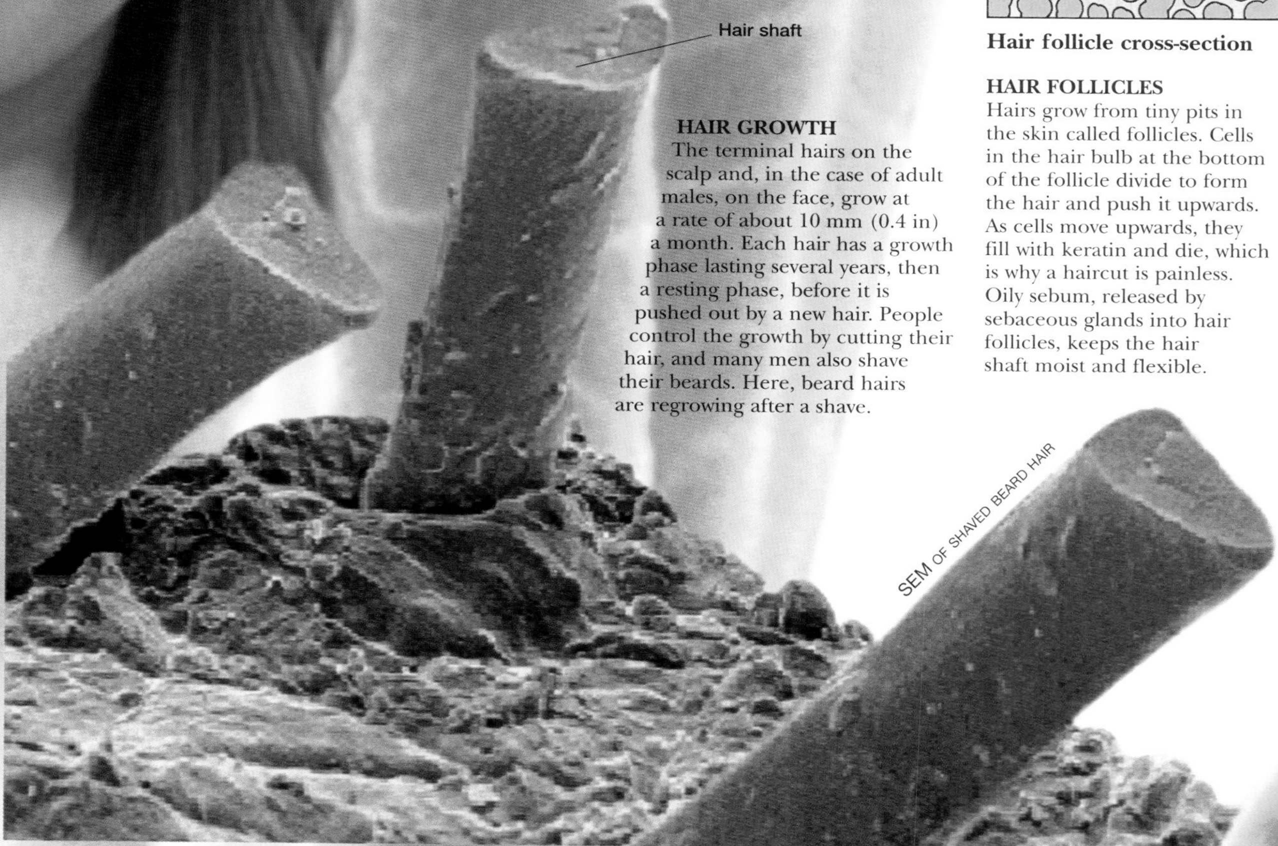

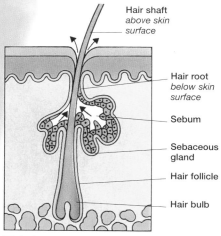

Hair shaft
above skin surface

Hair root
below skin surface

Sebum

Sebaceous gland

Hair follicle

Hair bulb

Hair follicle cross-section

HAIR GROWTH

The terminal hairs on the scalp and, in the case of adult males, on the face, grow at a rate of about 10 mm (0.4 in) a month. Each hair has a growth phase lasting several years, then a resting phase, before it is pushed out by a new hair. People control the growth by cutting their hair, and many men also shave their beards. Here, beard hairs are regrowing after a shave.

Hair shaft

SEM OF SHAVED BEARD HAIR

HAIR FOLLICLES

Hairs grow from tiny pits in the skin called follicles. Cells in the hair bulb at the bottom of the follicle divide to form the hair and push it upwards. As cells move upwards, they fill with keratin and die, which is why a haircut is painless. Oily sebum, released by sebaceous glands into hair follicles, keeps the hair shaft moist and flexible.

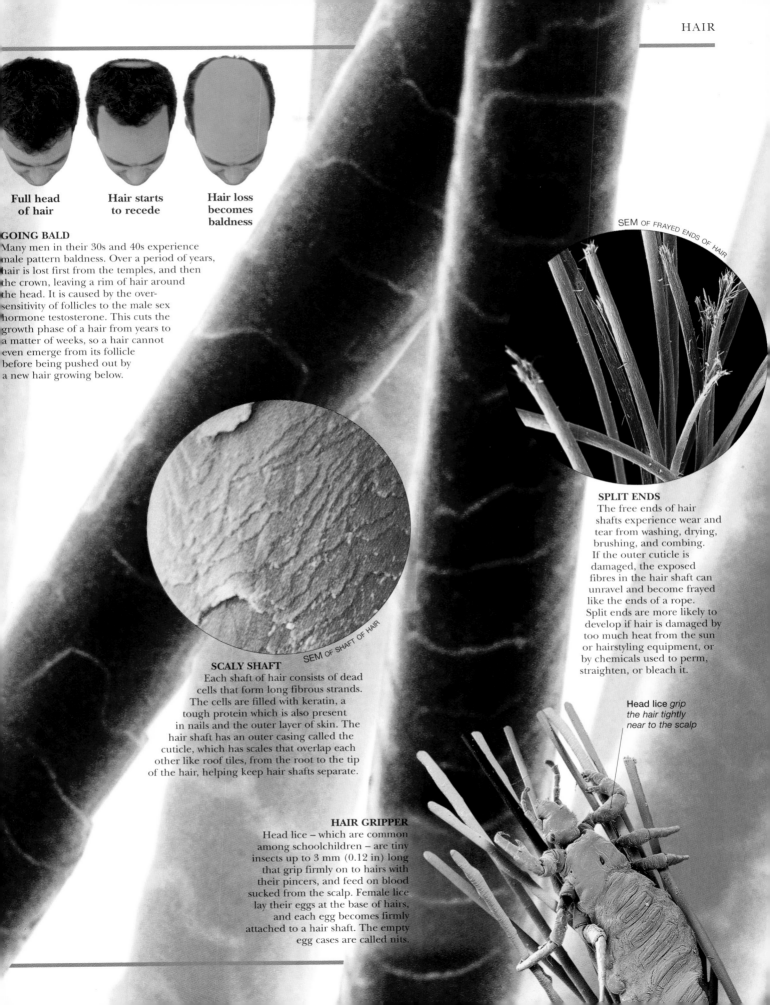

Full head of hair

Hair starts to recede

Hair loss becomes baldness

GOING BALD

Many men in their 30s and 40s experience male pattern baldness. Over a period of years, hair is lost first from the temples, and then the crown, leaving a rim of hair around the head. It is caused by the over-sensitivity of follicles to the male sex hormone testosterone. This cuts the growth phase of a hair from years to a matter of weeks, so a hair cannot even emerge from its follicle before being pushed out by a new hair growing below.

SEM OF FRAYED ENDS OF HAIR

SPLIT ENDS

The free ends of hair shafts experience wear and tear from washing, drying, brushing, and combing. If the outer cuticle is damaged, the exposed fibres in the hair shaft can unravel and become frayed like the ends of a rope. Split ends are more likely to develop if hair is damaged by too much heat from the sun or hairstyling equipment, or by chemicals used to perm, straighten, or bleach it.

SCALY SHAFT

Each shaft of hair consists of dead cells that form long fibrous strands. The cells are filled with keratin, a tough protein which is also present in nails and the outer layer of skin. The hair shaft has an outer casing called the cuticle, which has scales that overlap each other like roof tiles, from the root to the tip of the hair, helping keep hair shafts separate.

SEM OF SHAFT OF HAIR

Head lice grip the hair tightly near to the scalp

HAIR GRIPPER

Head lice – which are common among schoolchildren – are tiny insects up to 3 mm (0.12 in) long that grip firmly on to hairs with their pincers, and feed on blood sucked from the scalp. Female lice lay their eggs at the base of hairs, and each egg becomes firmly attached to a hair shaft. The empty egg cases are called nits.

SKELETAL system

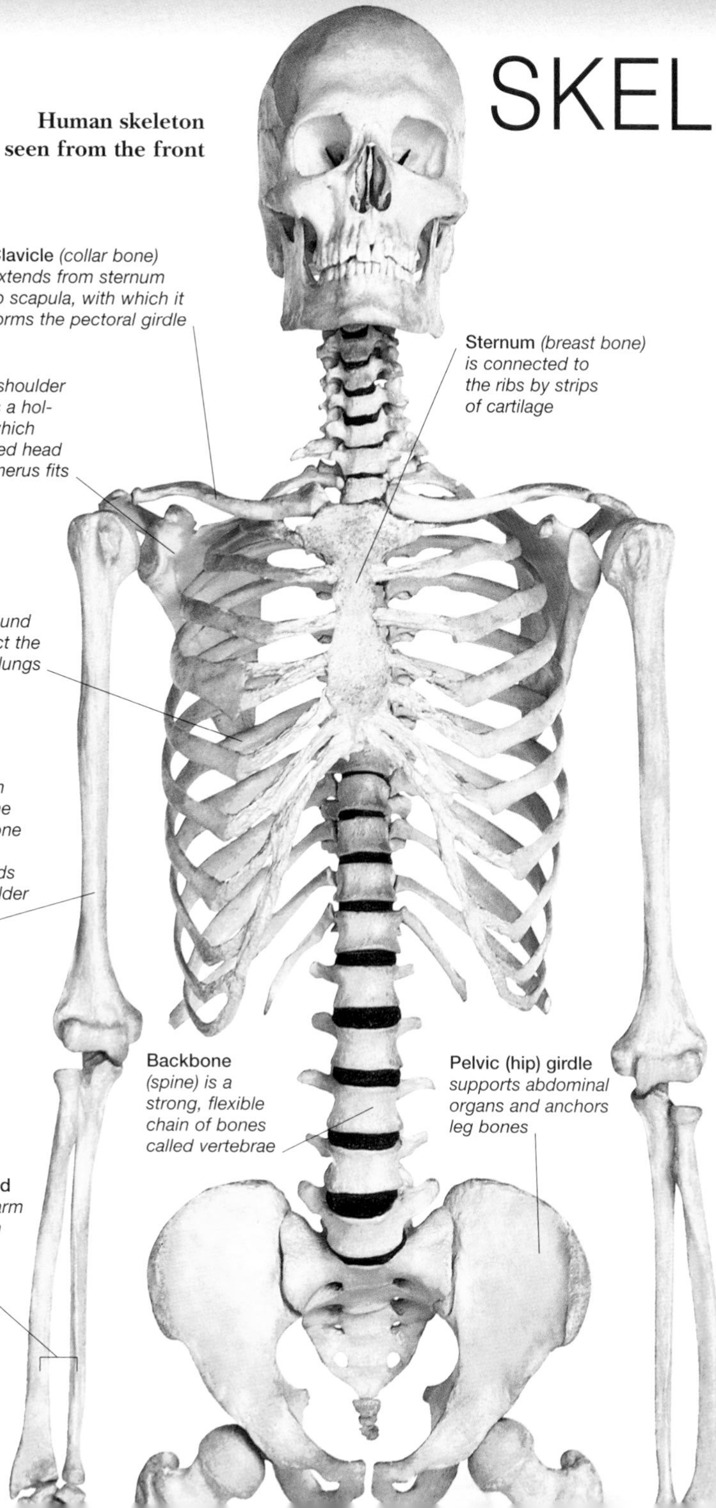

THE STRONG INNER framework of the body is formed by the skeletal system. Made up of separate bones that are linked together at joints, the skeletal system not only gives the body its shape, but also provides anchorage for the muscles that move it. It supports and protects vital organs, such as the brain, heart, and lungs. Although bones themselves are rigid, they are linked by joints that give the skeleton a great deal of flexibility. Before birth, the skeleton is made mostly of cartilage, and as a result is less rigid. As the body grows, cartilage is gradually replaced by bone, though some remains in the joints and also in the nose and ears.

LIVING FRAMEWORK

The skeleton is made up of 206 separate bones, which differ in shape, size, and name. The skull, backbone, ribs, and sternum form the central part of the skeleton. The bones of the arms and legs hang symmetrically on either side, attached respectively by the pectoral girdle (clavicle and scapula) and the pelvic girdle. Bones are not dry and lifeless, as may be thought, but are active organs containing living cells surrounded by protein fibres and mineral crystals. They act as mineral stores and are constantly exchanging calcium with blood. Inside some bones, marrow produces red and white blood cells.

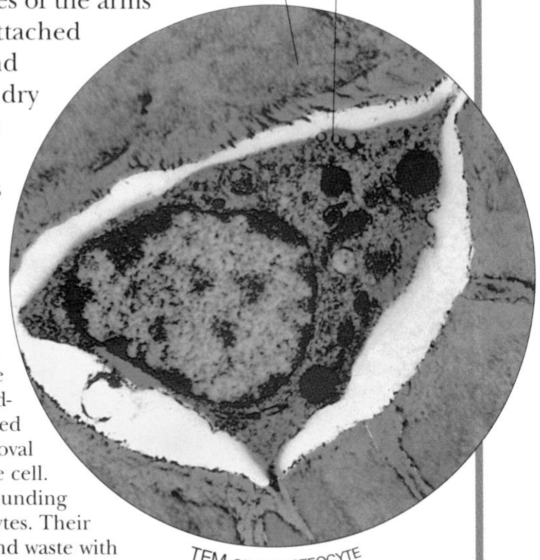

BONE CELL

This highly magnified image shows a mature bone cell called an osteocyte. It lies within a fluid-filled space called a lacuna, which is surrounded by compact bone. The cell has a large oval nucleus, which is visible in the lower part of the cell. Osteocytes send thin branches out into the surrounding bone to link with the branches of other osteocytes. Their role is to maintain bone, exchanging nutrients and waste with the blood. Other types of bone cell, called osteoblasts and osteoclasts, make bone and break it down respectively.

Compact bone

Osteocyte

TEM OF AN OSTEOCYTE

Human skeleton seen from the front

Clavicle (collar bone) extends from sternum to scapula, with it forms the pectoral girdle

Scapula (shoulder blade) has a hollow into which the rounded head of the humerus fits

Ribs surround and protect the heart and lungs

Humerus (upper arm bone) is the longest bone in the arm and extends from shoulder to elbow

Radius and ulna (forearm bones) run from the elbow to the wrist

Sternum (breast bone) is connected to the ribs by strips of cartilage

Backbone (spine) is a strong, flexible chain of bones called vertebrae

Pelvic (hip) girdle supports abdominal organs and anchors leg bones

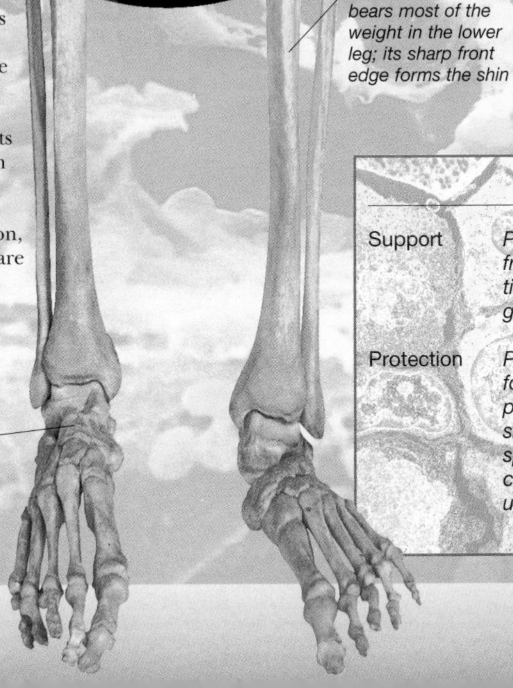

Metacarpal *is one of 27 hand bones that form the most flexible part of the skeleton*

Femur *(thigh bone) is the largest bone in the body. It has a rounded end that fits into the pelvic girdle; the other end has a wide, grooved surface that forms part of the knee*

Tibia *(shin bone) bears most of the weight in the lower leg; its sharp front edge forms the shin*

Tarsal *is one of 26 foot bones that supports the body during walking and standing*

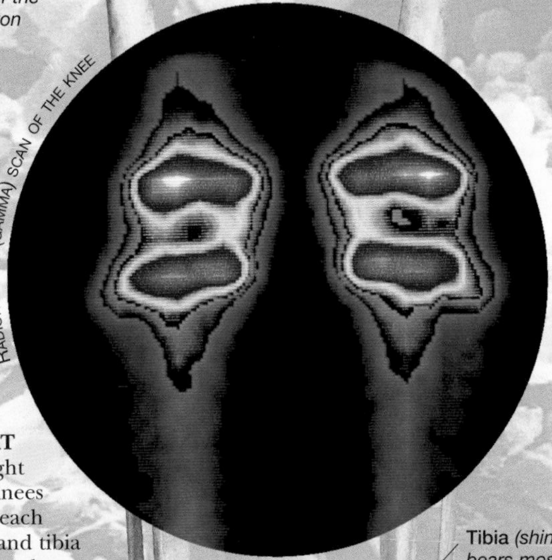

RADIONUCLIDE (GAMMA) SCAN OF THE KNEE

FLEXIBLE SUPPORT
The image on the right shows a scan of the knees – the joints between each femur (thigh bone) and tibia (shin bone). Joints are the parts of the skeleton where two or more bones meet. Held in place by strong bands of tissue called ligaments, they allow the bones to move and give the skeleton its flexibility. Each joint has its own range of movements, though most, like the knees, can move freely. In order to reduce friction, the ends of the bones at joints are covered with smooth cartilage.

18th-century engraving by Deuchar of Holbein's *Dance of Death*

SYMBOL OF DEATH
After death, once the flesh has rotted away, the bones are all that remain. Consequently, the skeleton has, for centuries, been used around the world as a symbol of death and disease. One example of this is the skull and crossbones which was used on the flags of pirates' ships to signal danger to others.

BONE SCAN
The image on the right is of a radionuclide (gamma) scan of the whole skeleton in a living person. Images such as these, which are produced in hospitals or clinics with special scanning equipment, can be very useful in medicine because they help doctors discover whether the bones are diseased. Doctors also use other tests, such as X-rays, magnetic resonance imaging (MRI), and ultrasound, to examine the skeletal system.

SKELETAL SYSTEM FUNCTIONS

Support	*Provides supportive framework for body tissues and organs; gives the body its shape.*	Movement	*Provides a strong yet light framework and anchorage for muscles; joints allow flexibility.*
Protection	*Provides protection for internal organs: ribs protect heart and lungs; skull protects the brain; spine protects the spinal cord; pelvis protects the uterus and bladder.*	Blood cell production	*Produces different types of blood cells in the red marrow of certain bones.*
		Mineral storage	*Acts as a reservoir for minerals, particularly calcium and phosphorus.*

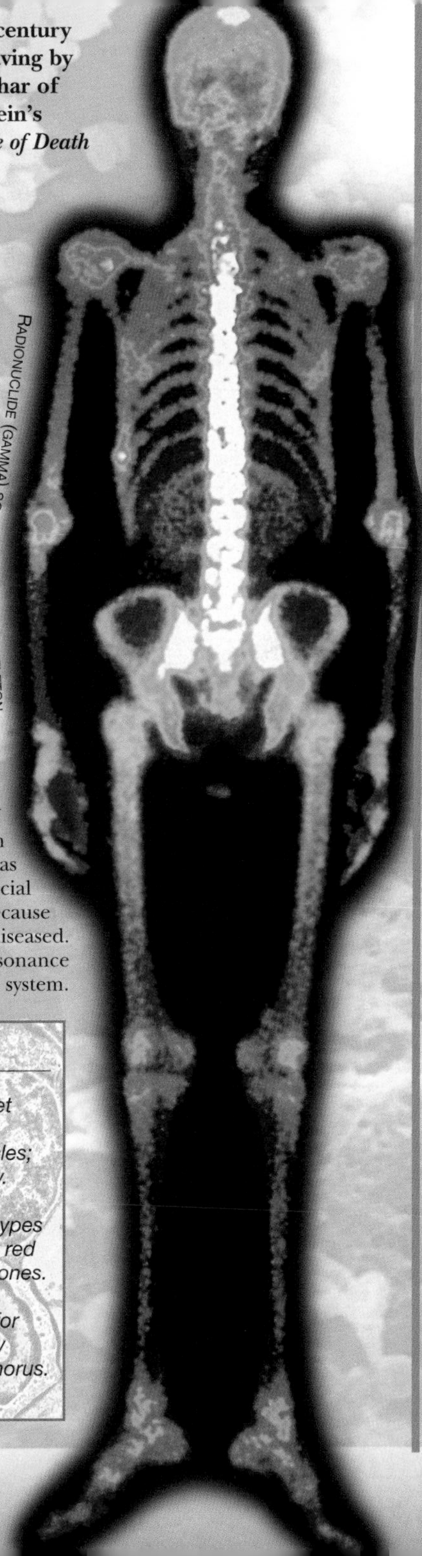

RADIONUCLIDE (GAMMA) SCAN OF A HEALTHY SKELETON

BACKGROUND: SEM OF BONE MARROW

Skull

T HE MOST COMPLEX PART of the skeleton, the skull shapes the head and face, protects the brain, and houses the special sense organs. It is made of 22 separate bones, 21 of which are locked together by immovable joints to form a structure of extraordinary strength. The only moveable skull bone is the mandible, or lower jaw. Skull bones are divided into two sets. The cranial bones form the domed upper part, or cranium, which surrounds, supports, and protects the brain and the organs of hearing. The facial bones form the framework of the face and jaw, and provide attachment sites for the muscles that produce facial expressions. Together, both cranial and facial bones form the orbits, or eye sockets, and the nasal cavity.

LOWER OPENING
At the base of the skull is a large circular hole – the foramen magnum (above, centre). The lowest part of the brain passes through this opening and continues downwards as the spinal cord. Other smaller holes are for the passage of nerves and blood vessels.

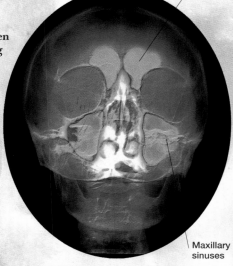

X-ray of the skull, seen from the front, showing two pairs of sinuses

Frontal sinuses

Maxillary sinuses

AIR SPACES
Some of the bones surrounding the nasal cavity contain hollow, air-filled spaces called sinuses. The sinuses lighten the skull's weight, and act as an echo chamber, giving a slight "ring" to the voice. They are lined with a moist membrane and connect through small openings with the inside of the nasal cavity.

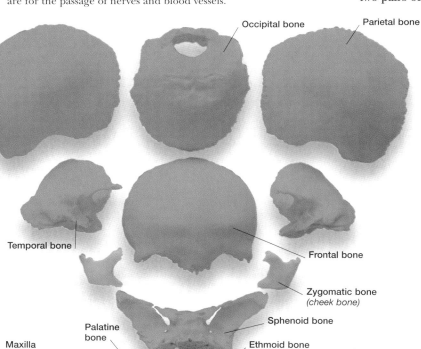

Occipital bone

Parietal bone

Temporal bone

Frontal bone

Zygomatic bone
(cheek bone)

Palatine bone

Sphenoid bone

Maxilla
(upper jaw bone)

Ethmoid bone

Inferior concha

Vomer

Bones of the skull

Cranial bones (8)

Facial bones (14)

Nasal bones

Mandible
(lower jaw)

COMPONENT PARTS
Eight bones form the cranium around the brain. The frontal bone is at the front, the two parietal bones form the sides and top, the occipital bone the back and – with the sphenoid – the base, the two temporal bones the side, and the ethmoid part of the nasal cavity. Each temporal bone has an opening to the inner parts of the ear which are encased within. The remaining 14 facial bones form the skeleton of the face. The zygomatic bones are the cheek bones. The palatine bones, nasal bones, inferior conchae, vomer, and lacrimal bones (not shown here) surround the nasal cavity. The maxillae (upper jaw bones) and the mandible (lower jaw bone) contain sockets for the teeth.

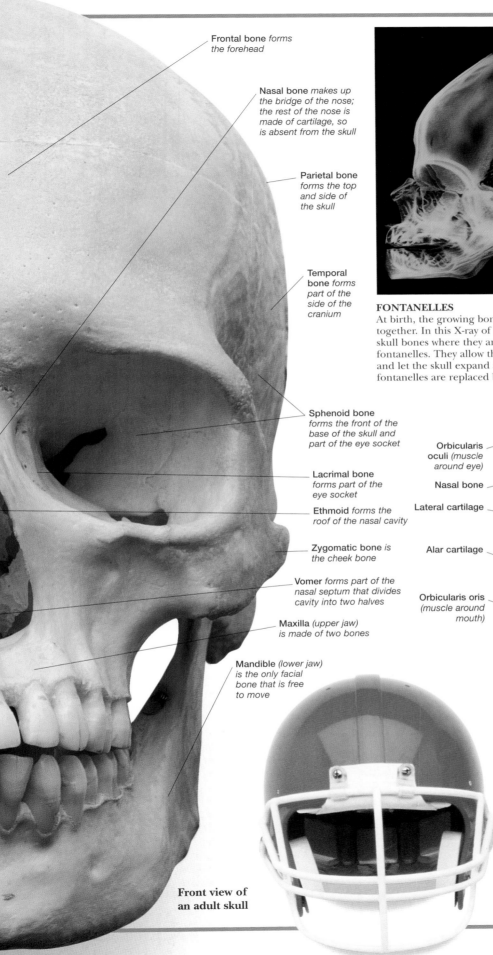

Frontal bone *forms the forehead*

Nasal bone *makes up the bridge of the nose; the rest of the nose is made of cartilage, so is absent from the skull*

Parietal bone *forms the top and side of the skull*

Temporal bone *forms part of the side of the cranium*

Sphenoid bone *forms the front of the base of the skull and part of the eye socket*

Lacrimal bone *forms part of the eye socket*

Ethmoid *forms the roof of the nasal cavity*

Zygomatic bone *is the cheek bone*

Vomer *forms part of the nasal septum that divides cavity into two halves*

Maxilla *(upper jaw) is made of two bones*

Mandible *(lower jaw) is the only facial bone that is free to move*

Front view of an adult skull

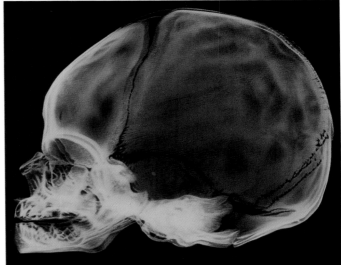

FONTANELLES

At birth, the growing bones of a baby's skull are not yet fixed firmly together. In this X-ray of a baby's skull, "gaps" are visible between skull bones where they are linked by areas of membranes called fontanelles. They allow the skull to be "squeezed" during birth and let the skull expand as the brain grows. By 12 to 18 months, fontanelles are replaced by bone.

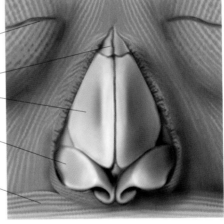

Orbicularis oculi *(muscle around eye)*

Nasal bone

Lateral cartilage

Alar cartilage

Orbicularis oris *(muscle around mouth)*

CARTILAGE FRAMEWORK

Cartilage is a tough, flexible tissue that is softer than bone. The nose (above) is shaped by several pieces. Lateral and alar cartilages form the sides of the nose, while septal cartilages form a central partition between the two nostrils. Cartilage is also present at the bone ends in mobile joints, at the front ends of the ribs, and in the ears.

EXTRA PROTECTION

Heavy blows to the head can be dangerous because they may damage the brain. Helmets provide extra protection against the risk of head injuries during hazardous activities, and have been worn for centuries by soldiers as part of their armour. They are also worn by cyclists and motor cyclists, by people participating in sports such as cricket and American football, and by those working on building sites or in unsafe environments.

Backbone and ribs

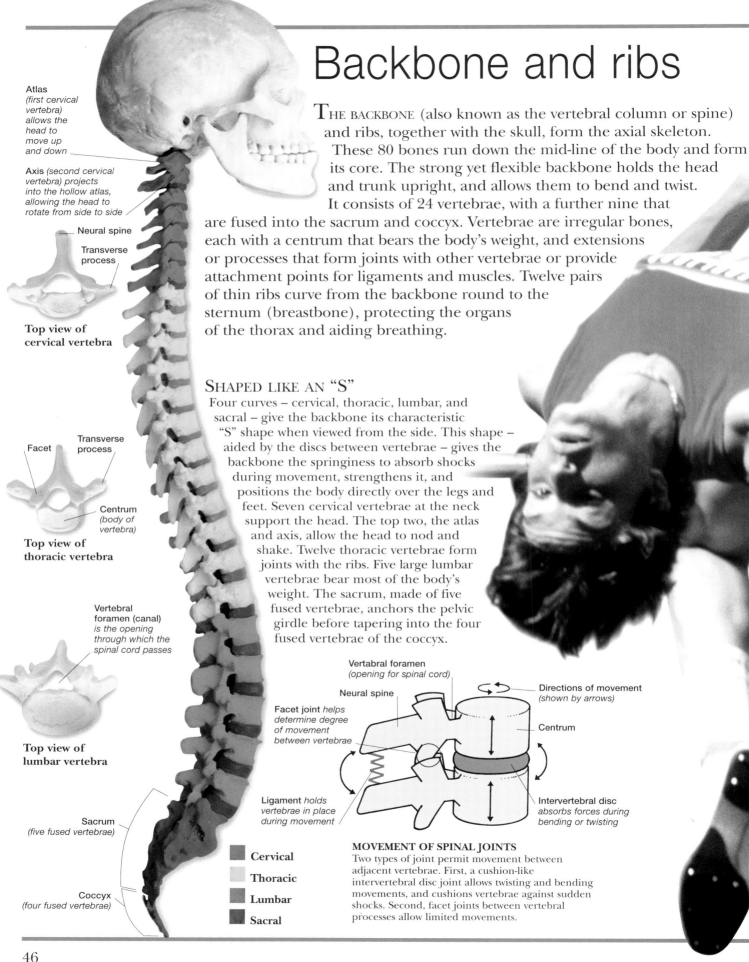

THE BACKBONE (also known as the vertebral column or spine) and ribs, together with the skull, form the axial skeleton. These 80 bones run down the mid-line of the body and form its core. The strong yet flexible backbone holds the head and trunk upright, and allows them to bend and twist. It consists of 24 vertebrae, with a further nine that are fused into the sacrum and coccyx. Vertebrae are irregular bones, each with a centrum that bears the body's weight, and extensions or processes that form joints with other vertebrae or provide attachment points for ligaments and muscles. Twelve pairs of thin ribs curve from the backbone round to the sternum (breastbone), protecting the organs of the thorax and aiding breathing.

SHAPED LIKE AN "S"

Four curves – cervical, thoracic, lumbar, and sacral – give the backbone its characteristic "S" shape when viewed from the side. This shape – aided by the discs between vertebrae – gives the backbone the springiness to absorb shocks during movement, strengthens it, and positions the body directly over the legs and feet. Seven cervical vertebrae at the neck support the head. The top two, the atlas and axis, allow the head to nod and shake. Twelve thoracic vertebrae form joints with the ribs. Five large lumbar vertebrae bear most of the body's weight. The sacrum, made of five fused vertebrae, anchors the pelvic girdle before tapering into the four fused vertebrae of the coccyx.

Atlas (first cervical vertebra) allows the head to move up and down

Axis (second cervical vertebra) projects into the hollow atlas, allowing the head to rotate from side to side

Neural spine

Transverse process

Top view of cervical vertebra

Facet

Transverse process

Centrum (body of vertebra)

Top view of thoracic vertebra

Vertebral foramen (canal) is the opening through which the spinal cord passes

Top view of lumbar vertebra

Sacrum (five fused vertebrae)

Coccyx (four fused vertebrae)

Cervical

Thoracic

Lumbar

Sacral

Vertabral foramen (opening for spinal cord)

Neural spine

Facet joint helps determine degree of movement between vertebrae

Ligament holds vertebrae in place during movement

Directions of movement (shown by arrows)

Centrum

Intervertebral disc absorbs forces during bending or twisting

MOVEMENT OF SPINAL JOINTS

Two types of joint permit movement between adjacent vertebrae. First, a cushion-like intervertebral disc joint allows twisting and bending movements, and cushions vertebrae against sudden shocks. Second, facet joints between vertebral processes allow limited movements.

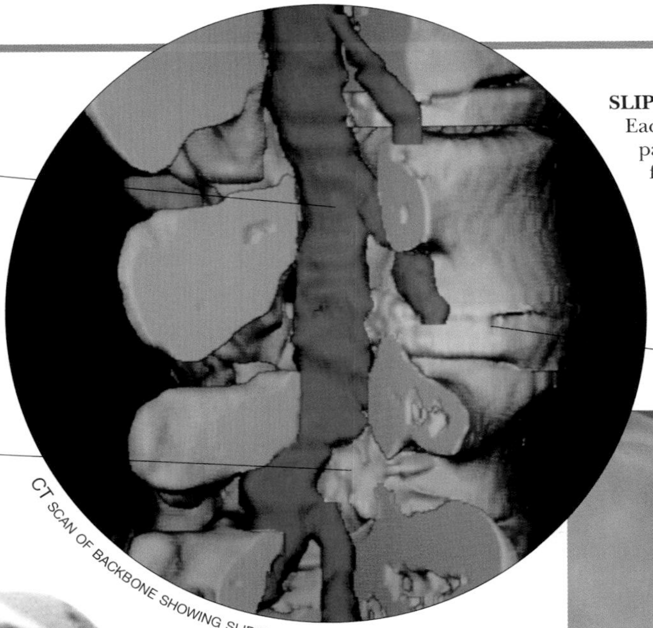

Spinal cord
*causes pain
and numbness
if compressed*

"Slipped disc"
*pressing against
the spinal cord*

CT SCAN OF BACKBONE SHOWING SLIPPED DISC

SLIPPED DISC

Each intervertebral disc between the vertebrae consists of a pad of fibrous cartilage with a jelly-like centre. Sometimes the fibrous coat breaks open, and part of the disc's core protrudes. If this happens, the core may put pressure on a spinal nerve or, as seen in this CT scan, the core (yellow) actually presses on the spinal cord (blue). This disc prolapse – known more commonly as a slipped disc – causes back pain, and may also cause weakness and pain in the arms and legs.

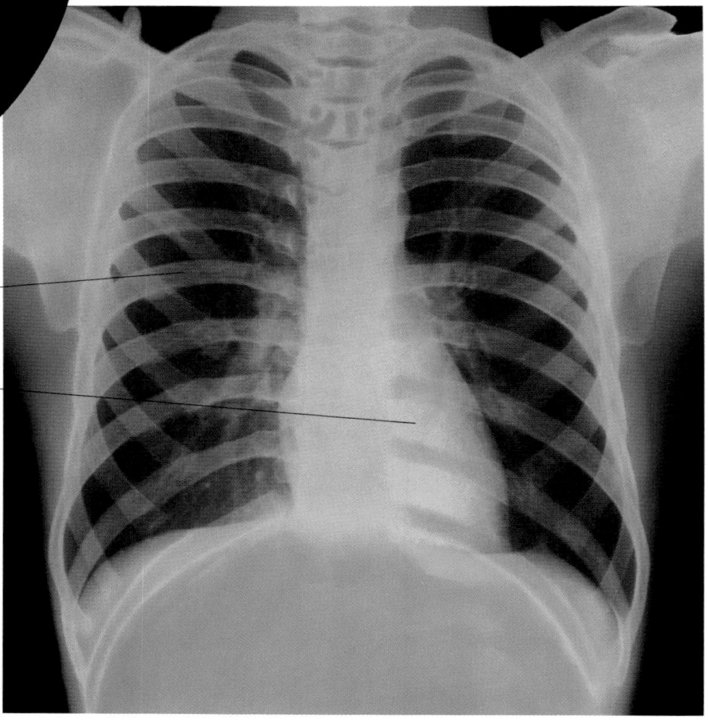

Intervertebral disc
in correct position

Ribs surrounding lungs

Heart

CHEST X-RAY OF AN 11-YEAR-OLD BOY

ORGAN PROTECTOR

This chest X-ray shows the cage-like structure formed by the ribs that protects the heart, lungs, and the organs of the upper abdomen. The 12 pairs of flat, curved bones extend from the backbone, where the rear end of each rib forms a joint with one of the thoracic vertebrae, around the wall of the thorax to meet the sternum at the front. Flexible costal cartilage connects the upper ten ribs to the sternum. The up and down movements of the ribcage during breathing move air in and out of the lungs.

SPINE FLEXIBILITY

The joints between neighbouring vertebrae that make up the bony chain running down the back allow only limited movement. But, added together, these small movements make the backbone as a whole very flexible. The backbone can bend forwards and backwards, and from side to side, and permits rotation with the body twisting on its axis. The body can bend further forwards (flexion) than it can backwards (extension) because the shape of vertebrae limits backward movement. Of all the vertebrae, the cervical vertebrae allow the greatest flexibility, a feature that becomes obvious in human neck movements.

This high-jumper demonstrates the flexibility of the spine

Limbs and girdles

WHENEVER SOMEONE jumps in the air, kicks a ball, swings a tennis racket, or writes a letter, he or she is making full use of their limbs and girdles. The limbs are the arms and legs, while the girdles are rings of bones that attach each limb pair to the axial, or central, part of the skeleton. Arm and leg bones share the same fundamental structure. Each has the same number of bones organized into three major segments – a single upper bone, a pair of lower bones, and a collection of small hand or foot bones – linked by freely movable joints. The pectoral (shoulder) girdle, consisting on each side of a scapula (shoulder blade) and a clavicle (collar bone), forms the link between the axial skeleton and the upper arm bones, while the pelvic (hip) girdle does the same job for the upper leg bones. The pectoral girdle is the weaker but more mobile of the two, and the pelvic is stronger but more rigid. Together, limbs and girdles make up that section of the bony framework called the appendicular skeleton.

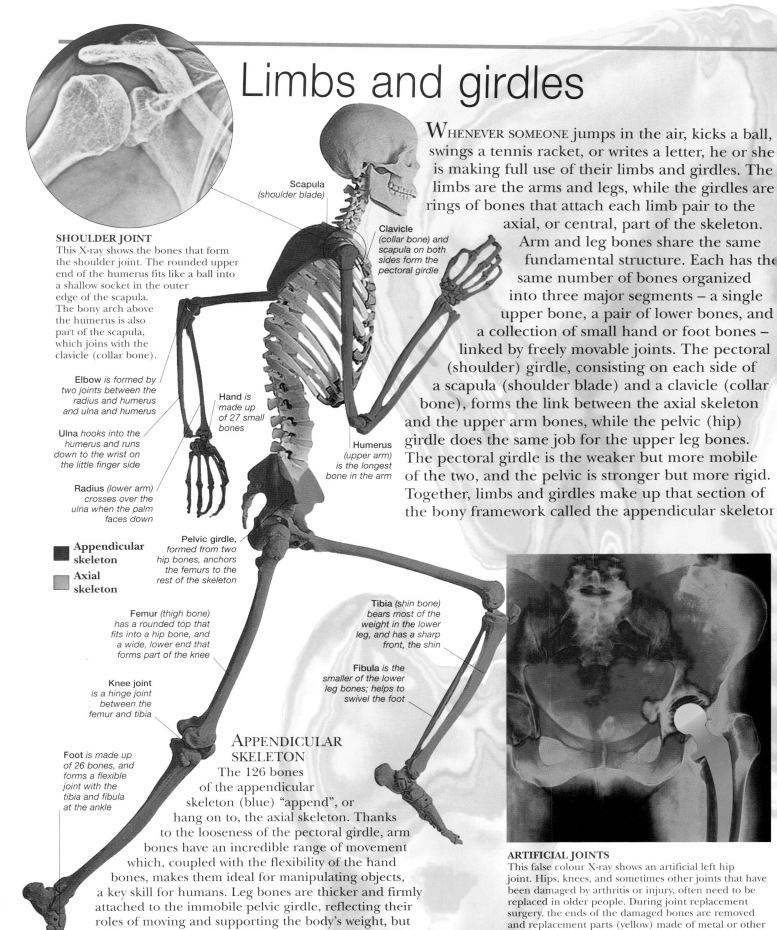

Scapula *(shoulder blade)*

Clavicle *(collar bone) and scapula on both sides form the pectoral girdle*

SHOULDER JOINT
This X-ray shows the bones that form the shoulder joint. The rounded upper end of the humerus fits like a ball into a shallow socket in the outer edge of the scapula. The bony arch above the humerus is also part of the scapula, which joins with the clavicle (collar bone).

Elbow *is formed by two joints between the radius and humerus and ulna and humerus*

Ulna *hooks into the humerus and runs down to the wrist on the little finger side*

Radius *(lower arm) crosses over the ulna when the palm faces down*

Hand *is made up of 27 small bones*

Humerus *(upper arm) is the longest bone in the arm*

■ **Appendicular skeleton**
□ **Axial skeleton**

Pelvic girdle, *formed from two hip bones, anchors the femurs to the rest of the skeleton*

Femur *(thigh bone) has a rounded top that fits into a hip bone, and a wide, lower end that forms part of the knee*

Knee joint *is a hinge joint between the femur and tibia*

Foot *is made up of 26 bones, and forms a flexible joint with the tibia and fibula at the ankle*

Tibia *(shin bone) bears most of the weight in the lower leg, and has a sharp front, the shin*

Fibula *is the smaller of the lower leg bones; helps to swivel the foot*

APPENDICULAR SKELETON
The 126 bones of the appendicular skeleton (blue) "append", or hang on to, the axial skeleton. Thanks to the looseness of the pectoral girdle, arm bones have an incredible range of movement which, coupled with the flexibility of the hand bones, makes them ideal for manipulating objects, a key skill for humans. Leg bones are thicker and firmly attached to the immobile pelvic girdle, reflecting their roles of moving and supporting the body's weight, but reducing their overall mobility compared with arm bones.

ARTIFICIAL JOINTS
This false colour X-ray shows an artificial left hip joint. Hips, knees, and sometimes other joints that have been damaged by arthritis or injury, often need to be replaced in older people. During joint replacement surgery, the ends of the damaged bones are removed and replacement parts (yellow) made of metal or other synthetic materials are inserted instead.

Anatomy, the study of the structure of the human body, has been a subject of great interest since ancient times. In the past, it was made difficult because many cultures forbade the dissection, or cutting up, of dead bodies to look inside. For that reason, the easiest way to study anatomy was to look at the bones, the only part of the body to remain long after death. This illustration from 1363 shows a lesson in skeletal anatomy given by the French anatomist Guy de Chauliac.

Guy de Chauliac (right) giving an anatomy lesson

**Head of femur
(actual size)**

**Stapes
(actual size)**

LARGE AND SMALL
The largest bone in the body is the femur. In an average adult it is about 46 cm (18 in) long and has a tube-shaped shaft, making it longer and stronger than any other bone. At the upper end, its rounded head fits into a socket in the pelvis at the hip joint. The lower end has a wide, grooved surface that forms a joint with the tibia in the knee. The smallest bone in the body is the stapes – one of three tiny bones, called ossicles, situated inside the ear – which is shaped like a stirrup, and measures an average of 5 mm (0.2 in) in length.

Hip bone – *the curved shape of these bones supports the abdominal organs*

Male pelvis

Pelvic inlet *is narrower in males*

Pubic symphysis *is the joint between hip bones at front of pelvic girdle*

Femur socket *is where the femur's rounded head fits into the pelvic girdle*

Sacrum, *part of the backbone, is attached by powerful ligaments to the pelvic girdle to form the pelvis*

PELVIC GIRDLE AND PELVIS
Two curved hip bones make up the pelvic girdle, the structure that forms the attachment point for the thigh bones, and which transmits weight downwards from the upper body. The hip bones meet at the front of the girdle, while at the rear they are firmly attached to the sacrum. Together, the pelvic girdle and sacrum form the pelvis, a bowl-shaped structure that supports and protects digestive, reproductive, and urinary organs. The opening in the centre of the pelvis – the pelvic inlet – is wider in females than in males, thus providing sufficient space for a baby's head to squeeze through during birth.

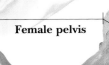

Female pelvis

Pelvic inlet *is wider in females*

Hands and feet

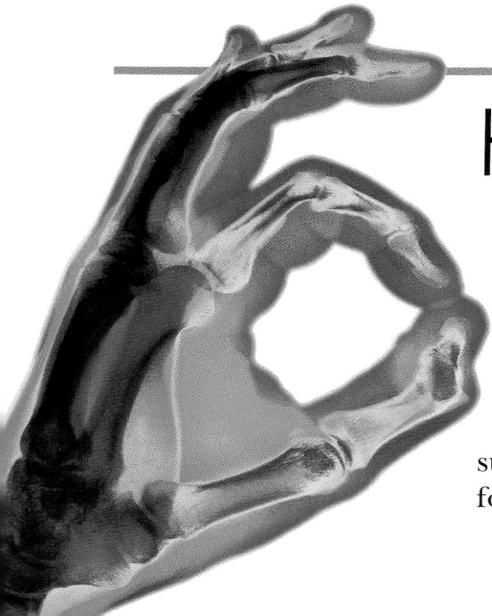

THE HINGED ATTACHMENTS at the lower ends of the limbs are called hands and feet. The hands are attached at the wrist joints, and the feet at the ankle joints. Although both hands and feet have five flexible digits, the hands are more versatile. They can perform an enormous range of movements. In particular, they can form a grip that allows them to hold and handle objects. The feet are mobile platforms for the body. They support its weight and also act as springboards to propel the body forward during walking, running, or jumping.

OPPOSABLE THUMB
The thumb is the most mobile digit. It can swing across the palm and turn towards the fingers, so that its tip touches the fingertips. This action is called opposition, and it enables the hand to form a grip. A precise grip between the thumb and forefinger, as shown in the above X-ray, is useful for handling small objects.

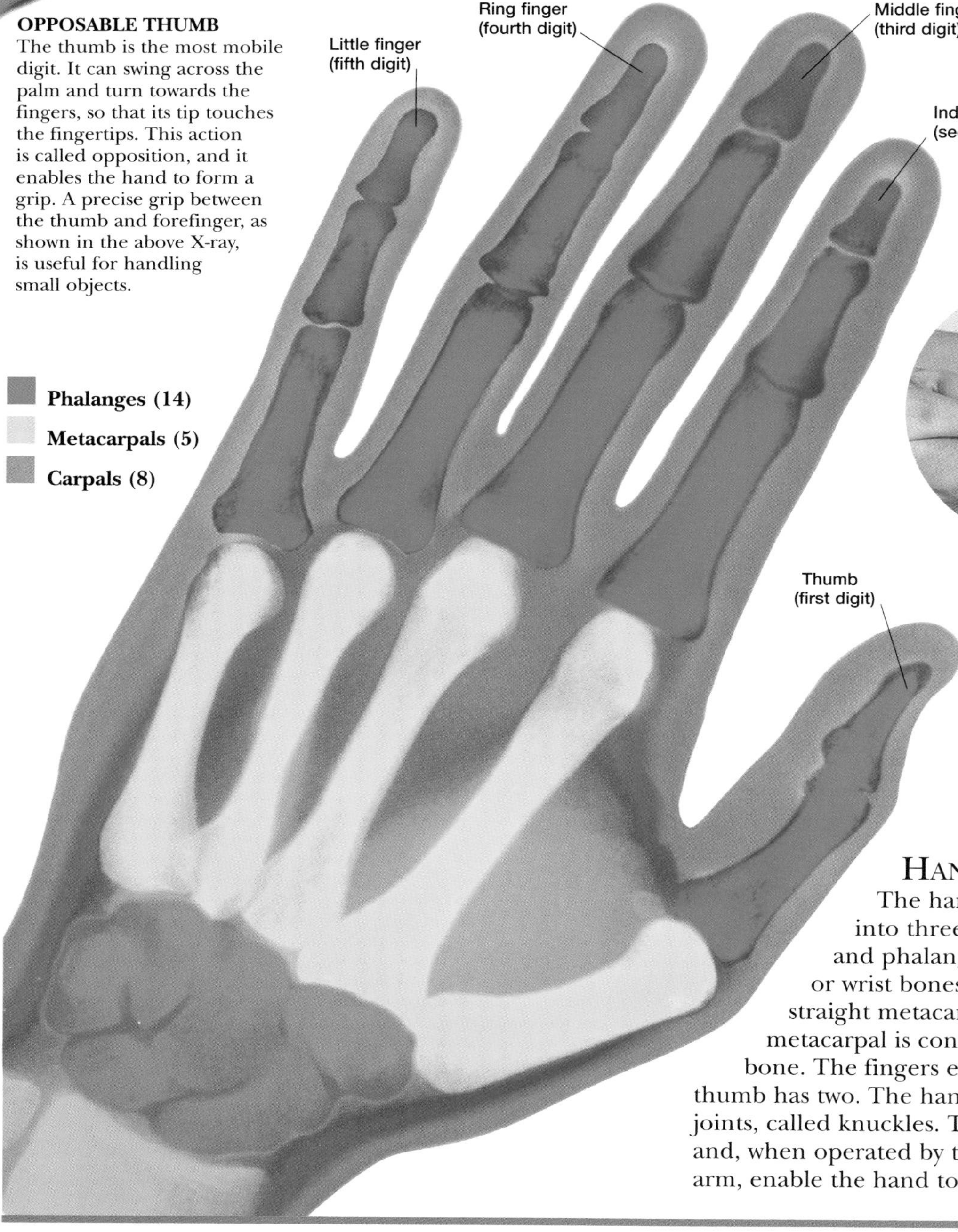

Little finger (fifth digit)

Ring finger (fourth digit)

Middle finger (third digit)

Index finger (second digit)

Thumb (first digit)

■ Phalanges (14)
■ Metacarpals (5)
■ Carpals (8)

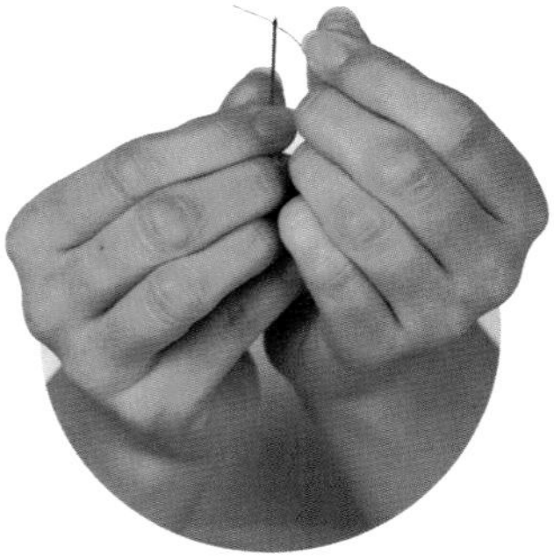

Threading a needle

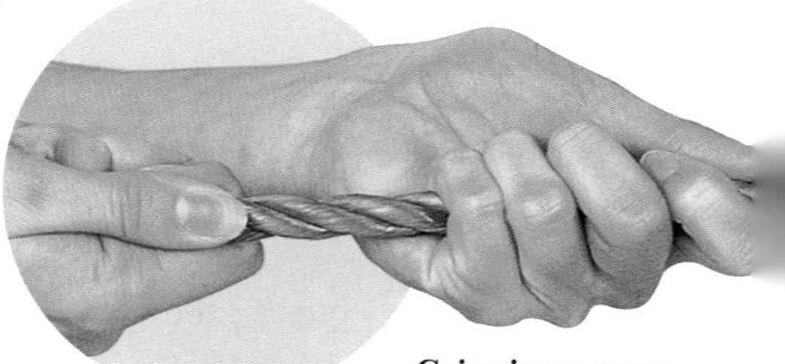

Gripping a rope

PRECISION AND POWER
The hand can grip precisely between thumb and fingertips for delicate tasks such as holding a pen or threading a needle. It can also make a more secure and powerful grip by wrapping thumb and fingers around an object, as when pulling a rope.

HANDS
The hand contains 27 bones divided into three groups: carpals, metacarpals, and phalanges. Eight of the bones are carpals, or wrist bones. Connected to the carpals are five straight metacarpals, which form the palm. Each metacarpal is connected to a phalanx, or finger bone. The fingers each have three phalanges and the thumb has two. The hand has a total of 14 finger or thumb joints, called knuckles. These give the hand great flexibility and, when operated by the muscles of the hand and lower arm, enable the hand to perform a multitude of tasks.

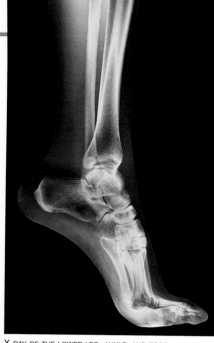

X-RAY OF THE LOWER LEG, ANKLE, AND FOOT

Great (big) toe (first digit)

Second toe (second digit)

Third toe (third digit)

Fourth toe (fourth digit)

Little toe (fifth digit)

JUMP OFF

The feet provide support and stability to prevent the body from falling over while standing or moving on flat or uneven surfaces. The feet are also strong flexible levers that push the body off the ground during walking, climbing, running, and jumping. This X-ray shows the bones of the foot, ankle joint, and lower leg as the foot pushes the body up. The toes are the only part that remain in contact with the ground.

FEET

Each foot contains 26 bones arranged in three groups: tarsals, metatarsals, and phalanges. There are seven tarsals, or ankle bones. The largest of these is the calcaneus, or heel bone, which projects backwards behind the ankle joint. The calcaneus is connected to the talus, which forms a joint with the bones of the lower leg. In front of the tarsals are five metatarsal bones, or sole bones, linked to the 14 phalanges, two of which are in the big toe, and three in each other toe. The feet are less flexible than the hands because they are bound together by strong ligaments.

Talus *forms a joint with the tibia and fibula at the ankle*

FLEXIBLE ARCH

Footprints made by bare feet reveal that the soles are not flat and that only part of the foot touches the ground. Flexible curves, called arches, raise part of the foot off the ground. The arches are held up by ligaments and by tendons pulled by their muscles. Arches help spread out body weight and provide a springiness that absorbs shocks during running and walking. Feet with reduced arches are called flat feet.

Calcaneus *(heel bone)*

Phalanges (14)

Metatarsals (5)

Tarsals (7)

BONES IN EVOLUTION

MUCH OF OUR KNOWLEDGE of our ancestry comes from the study of the human skeleton. Bones and teeth are the only parts of the body that form fossils. This usually occurs when human remains become buried in mud, and bone tissue is replaced by minerals. After death, bones become separated, so it is extremely rare to find more than scattered fossilized bones and skull fragments. But painstaking searches in east Africa, followed by careful measurements and comparisons of bones, have allowed palaeontologists to piece together the story of our evolution from our ape-like ancestors.

PROCONSUL AFRICANUS
Skeletons of ape-like *Proconsul africanus* (above), which lived about 20 million years ago, show long pelvic bones and arms typical of primates that walk using all four limbs. By 5 million years ago, hominids have strong leg bones and a broad pelvis, suggesting they walked on two legs.

ANCIENT FOOTPRINTS
This is the fossil footprint of *Australopithecus afarensis*, discovered at Laetoli, Tanzania, in 1972. The big toe (top right) hardly diverges from the rest of the foot, unlike that of primates such as chimpanzees that walk on all fours and use this toe like a thumb. This fossil demonstrates conclusively that these early humans walked upright on two legs.

DOWN FROM THE TREES

Humans belong to a group of mammals called primates, and share a common ancestor with lemurs, monkeys, and humans. The hominids, a family of upright-walking humans to which we belong, arose some 6 million years ago, at a time when the climate in Africa became drier and grasslands began to replace forests. As our ape-like ancestors spent more time foraging and hunting on the ground, a species that walked upright evolved. It seems likely that the ability to see approaching danger over tall grasses, the advantage of having hands free to manipulate objects, and the necessity to cool the body in a hot climate were all important factors in the transition from four legs to two.

EARLY FAMILY HISTORY

Fossil bones of ape-like *Proconsul africanus*, which lived 20 million years ago, provide palaeontologists with a point of reference for comparing more recent hominid fossils. *Proconsul* had the arched backbone, long pelvic bones, and long fore limbs that are typical of a primate that walks on all four limbs, with gripping hands and feet for climbing. The earliest hominids that walked upright were the Australopithecines (meaning "southern ape"). In 1974, American Donald Johanson (b. 1943) found the fossilized partial skeleton of *Australopithecus afarensis* in the Afar region of Ethiopia.

Clavicle

Large, projecting jaw

Humerus

Vertebra

Carpal

Femur *(thigh bone)*

Knee joint

Tibia *(upper piece)*

AUSTRALOPITHECUS AFARENSIS
About 40 per cent of the skeleton of "Lucy" has been found. She belonged to the species *Australopithecus afarensis*, the first known primate to have walked upright. Lucy was about 143 cm (4 ft 9 in) tall and is named after the Beatles' song "Lucy in the Sky with Diamonds", which was popular among the palaeontologists who found her.

HOMO HABILIS

Fossil bones of *Homo habilis* (meaning "handy man") have been found among the stone tools that he used. This species was only about 135 cm (4 ft 6 in) tall and had a more rounded head, narrower jaw, and longer face than *Australopithecus*.

Homo habilis **used simple tools**

"Lucy", as the specimen became known, was 3 million years old. She had the curved, S-shaped spine and short pelvis that is typical of a primate that walks upright. Other key features of upright-walking primates include big toes aligned with all of the others, and feet with pronounced arches. In 1972, Mary Leakey (1913–96) found 3.5-million-year-old footprints at Laetoli in Tanzania that had been left by feet just like this, preserved in hardened volcanic ash. These provide the earliest direct evidence of primates walking upright.

HUMANS APPEAR

Fossil bones of *Homo habilis*, the earliest member of our own genus, were found by Mary and Louis Leakey (1903–72) in 1960, in the Olduvai Gorge of Tanzania. This species lived about 2 million years ago, and it had a small cranium, slightly protruding face, and long, thick leg bones. Fossil bones of *Homo erectus*, which evolved about 1.8 million years ago, have been found throughout

HOMO ERECTUS

Homo erectus was taller than *Homo habilis* and colonized Europe and Asia. Chinese fossils dating from 360,000 years ago have been found in caves amongst ash, charcoal, and charred bones – showing that this species had learned to use fire.

Europe and Asia, suggesting that this was the first species to have travelled out of Africa and colonized other continents. Fossil bones show that Neanderthals then evolved as a distinct species about 250,000 years ago, and modern humans evolved alongside them until Neanderthals became extinct, about 35,000 years ago.

Eyebrow ridges *are very distinctive in Neanderthals*

Skull of *Homo neanderthalensis*

HOMO NEANDERTHALENSIS

Studies of fossil hominid skulls show that skull size and brain volume have increased, teeth have become smaller, jaws less jutting, eyebrow ridges less prominent, and the chin more pronounced. Neanderthals had broad skulls and pronounced eyebrow ridges, but could easily pass unnoticed in the street today.

PALAEONTOLOGIST AT WORK

Palaeontologists rarely find more than fragments of skeletons. These are collected, photographed, carefully measured, and used to reconstruct the appearance of our earliest ancestors. Computer technology is often used to analyse and reconstruct skeletons.

Bone structure

BONES ARE REMARKABLE pieces of engineering, combining lightness with a strength that rivals steel. They are also slightly springy, which helps them to withstand sudden jolts and knocks. This unusual mix of features comes from their underlying structure, which consists of living cells and protein fibres, wrapped around layers of hard mineral salts. Bones are at their densest and toughest near the outside, where they bear the most stress. On the inside they have a sponge-like design that helps to reduce weight. Long bones – like the ones in the arms and the legs – also have a large central cavity. This is packed with marrow, a substance that stores fat and makes blood cells.

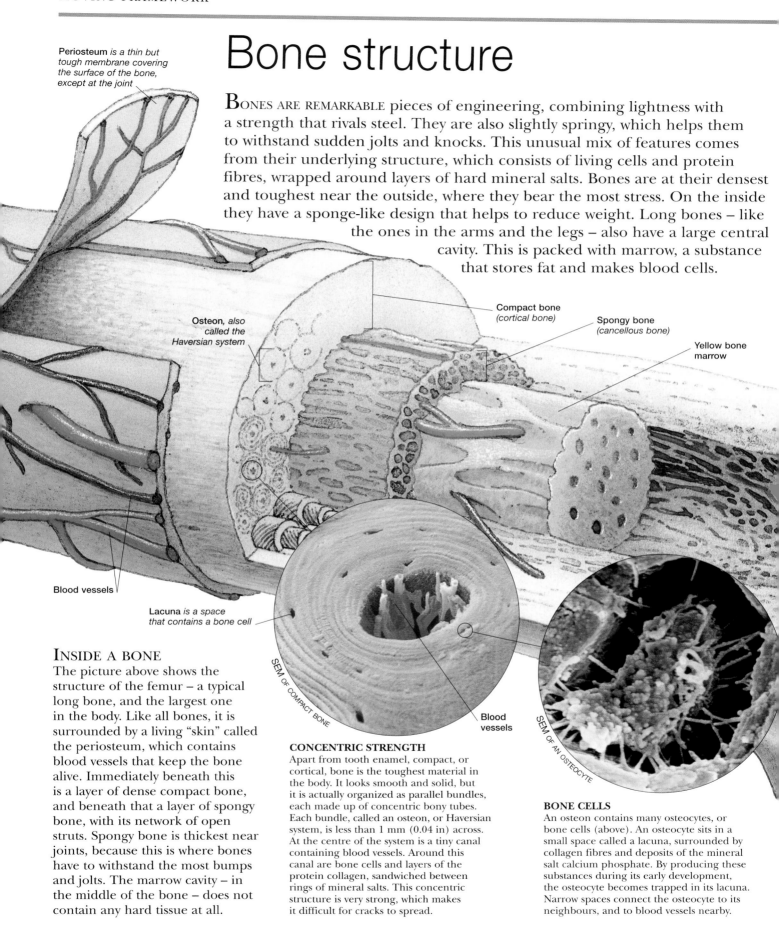

Periosteum is a thin but tough membrane covering the surface of the bone, except at the joint

Osteon, also called the Haversian system

Compact bone *(cortical bone)*

Spongy bone *(cancellous bone)*

Yellow bone marrow

Blood vessels

Lacuna is a space that contains a bone cell

SEM OF COMPACT BONE

SEM OF AN OSTEOCYTE

Blood vessels

INSIDE A BONE

The picture above shows the structure of the femur – a typical long bone, and the largest one in the body. Like all bones, it is surrounded by a living "skin" called the periosteum, which contains blood vessels that keep the bone alive. Immediately beneath this is a layer of dense compact bone, and beneath that a layer of spongy bone, with its network of open struts. Spongy bone is thickest near joints, because this is where bones have to withstand the most bumps and jolts. The marrow cavity – in the middle of the bone – does not contain any hard tissue at all.

CONCENTRIC STRENGTH
Apart from tooth enamel, compact, or cortical, bone is the toughest material in the body. It looks smooth and solid, but it is actually organized as parallel bundles, each made up of concentric bony tubes. Each bundle, called an osteon, or Haversian system, is less than 1 mm (0.04 in) across. At the centre of the system is a tiny canal containing blood vessels. Around this canal are bone cells and layers of the protein collagen, sandwiched between rings of mineral salts. This concentric structure is very strong, which makes it difficult for cracks to spread.

BONE CELLS
An osteon contains many osteocytes, or bone cells (above). An osteocyte sits in a small space called a lacuna, surrounded by collagen fibres and deposits of the mineral salt calcium phosphate. By producing these substances during its early development, the osteocyte becomes trapped in its lacuna. Narrow spaces connect the osteocyte to its neighbours, and to blood vessels nearby.

OSTEOPOROSIS

As people grow older, they can lose calcium and protein in their bones, causing the bones to become lighter and more likely to break. This condition, known as osteoporosis, affects both women and men, though women are particularly at risk because of changes in their hormone levels following the menopause (see p. 242). The photograph on the left shows spongy bone that has been weakened in this way. Compared with healthy spongy bone, shown below, its struts are full of spaces, or pores.

SEM OF SPONGY BONE AFFECTED BY OSTEOPOROSIS

Compact bone
(cortical bone)

Spongy bone
(cancellous bone)

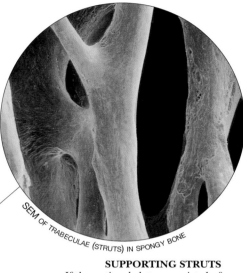

SEM OF TRABECULAE (STRUTS) IN SPONGY BONE

SUPPORTING STRUTS

If the entire skeleton consisted of compact bone, it would be much too heavy to move. Fortunately, spongy, or cancellous, bone helps to reduce its weight. Despite its name, it is not really spongy. Instead, it is made up of a network of rigid struts, which takes some of the load that the bone has to bear. These struts are separated by a maze of tiny spaces, which are filled with bone marrow.

BONE MARROW

Marrow is an important tissue because it is responsible for making blood cells. At birth, all bones contain blood-cell producing red marrow, but, during the teenage years, some of it is replaced by yellow marrow, which consists mainly of fat. In adults, red marrow is found mainly in spongy bone. Yellow marrow is concentrated in the central cavity of long bones, where it acts mainly as a storage tissue.

SEM OF RED BONE MARROW

**Structure of a femur
(thigh bone)**

Growth and repair

Unlike scaffolding or steel girders, bones change all the time. They grow in step with the rest of the body, can repair minor fractures, and even some major breaks. Even when they have reached adult size, they are constantly but invisibly renewed. How do they do it? The answer lies in the cells that all bones contain. Some of these cells, called osteoblasts, deposit protein and mineral salts, forming new bone. At the same time, other cells, called osteoclasts, do the reverse. By adjusting the balance between these two processes, the body makes bones grow. It also ensures that bones keep strong, by renewing or "remodelling" parts that get the most wear and tear.

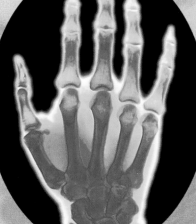

Bone (dark areas) replaces cartilage as the baby grows

Joints are one of the last areas to be ossified

Skeleton of a 14-week-old fetus

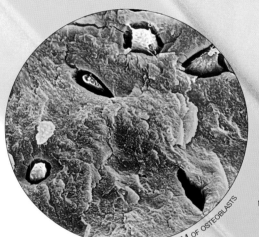

SEM OF OSTEOBLASTS

EARLY DEVELOPMENT

When the skeleton first forms, it consists solely of cartilage. During a process called ossification, osteoblast cells lay down mineral salts in this framework, gradually turning most of it into bone. This image shows a fetus that is 14 weeks old. The dark areas show that its bones are partially ossified, but its joints still consist of cartilage. Ossification continues after birth, and it lasts throughout adult life. For example, the flap of cartilage below the breastbone, called the xiphoid process, often ossifies at the age of 40 or older.

BUILDING BONES

This close-up view shows active osteoblasts (bone-making cells) on the surface of a bone. They lie just underneath the bone's outer "skin" or periosteum, and build up the bone from the outside. As a bone grows, they gradually become embedded in the bone itself. Once this has happened, they retire from bone-making, and change into mature bone cells or osteocytes.

GROWING BONES

The X-rays below show how hand bones develop between infancy and early adulthood. At first, the bones are only partially ossified, and their ends consist of cartilage. But as the cartilage grows, the ossified zone also expands, until finally the adult bone is complete. When bones grow, most of the growth occurs at their ends. Existing bone tissue often has to be broken down so that the bone can reach its adult shape.

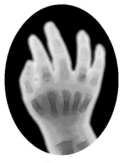

1 year

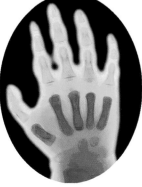

3 years

13 years

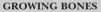

20 years

CARDIAC MUSCLE

The branched, striated fibres of cardiac muscle form an interconnected network in the heart wall. Cardiac muscle contracts automatically, with its own built-in rhythm, some 100,000 times each day. It cannot be controlled consciously. However, its rate of contraction is regulated by the autonomic nervous system, according to whether the body is active or at rest.

LM OF CARDIAC MUSCLE FIBRES

Skeletal muscle is covered by a connective tissue sheath called the epimysium

Tendon makes a secure attachment between the muscle and a bone

Bone is pulled by the tendon when a muscle contracts

Periosteum is the membrane that covers the bone's surface

BONE CONNECTORS

Skeletal muscles are anchored to bones – and sometimes to each other – by strong cords or sheets called tendons that enable muscles to pull bones without tearing. Tendons have enormous tensile strength (resistance to breaking). They are packed with fibres of the tough protein collagen, which are arranged in parallel bundles. The connective tissue sheath that encloses the muscle extends to the tendon. The tendon then extends to the bone, where its fibres penetrate the periosteum (surface membrane) and embed themselves firmly in the bone's outer layer, completing the connection between muscle and bone.

HEAT GENERATORS

By making use of heat generated inside it, the body maintains a constant internal temperature of 37°C (98.6°F). About 85 per cent of this heat is produced by muscle contraction. Of the energy converted by muscles, only 25 per cent causes movement. The remaining 75 per cent is "waste" energy, released as heat. This thermogram reveals how body heat is lost through the skin – lighter colours mark areas of greatest heat radiation.

MUSCULAR SYSTEM FUNCTIONS

Movement	Skeletal muscles produce a wide range of movements including running, picking up objects, and changing facial expressions. Cardiac muscle in the heart pumps blood to all body tissues. Smooth muscle produces movement in internal organs, such as pushing food along the intestines.	Joint stability	When they contract, some skeletal muscles help to stabilize highly mobile joints such as shoulders.
		Heat generation	Because they are not totally efficient, muscles generate heat as a by-product when they contract. Heat generation, or thermogenesis, is vitally important in maintaining normal body temperature. In cold conditions, the body uses involuntary contractions (shivering) to generate additional heat.
Posture maintenance	Certain skeletal muscles are kept in a partially contracted state to hold the body upright.		

Skeletal muscles

IN ANY DESCRIPTION of the muscular system, the skeletal muscles take centre stage. Making up nearly half of the body's mass, skeletal muscles can perform a wide range of movements, from blinking an eyelid to wielding a sledge-hammer. They are primarily attached to bones by tough, fibrous tendons. Typically, each muscle links two bones across a flexible joint so that muscle contraction, or shortening, either results in movement or assists in holding the body upright. A few muscles – such as those that produce facial expressions – work by tugging on the skin.

NAMING MUSCLES

Every skeletal muscle is given a Latin name according to one or more of its features, as described below. Some muscle names cover several features. The extensor carpi radialis longus, for example, extends (straightens) the wrist ("carpi"), lies close to the radius (lower arm) bone, and is longer than other wrist extensors.

MUSCLE FEATURES AND DESCRIPTIONS

Location Example: the frontalis runs over the frontal bone of the skull.

Relative size using terms such as maximus (largest), minimus (smallest), longus (long), and brevis (short). Example: the gluteus maximus is the biggest gluteal (buttock) muscle.

Shape Example: the two trapezius muscles form a trapezoid (four-sided) shape.

Action using terms such as flexor (bends a joint) and extensor (straightens a joint). Example: the flexor carpi ulnaris bends the hand at the wrist.

Origin and insertion Example: the sternocleidomastoid has origins (where bones do not move) on the breastbone – sternum – and collar bone – clavicle ("cleido") – and insertions (where bones do move) on the mastoid process of the skull's temporal bone.

Number of origins using terms such as biceps ("two heads").
Example: the biceps brachii (arm) has two origins on the scapula, or shoulder blade.

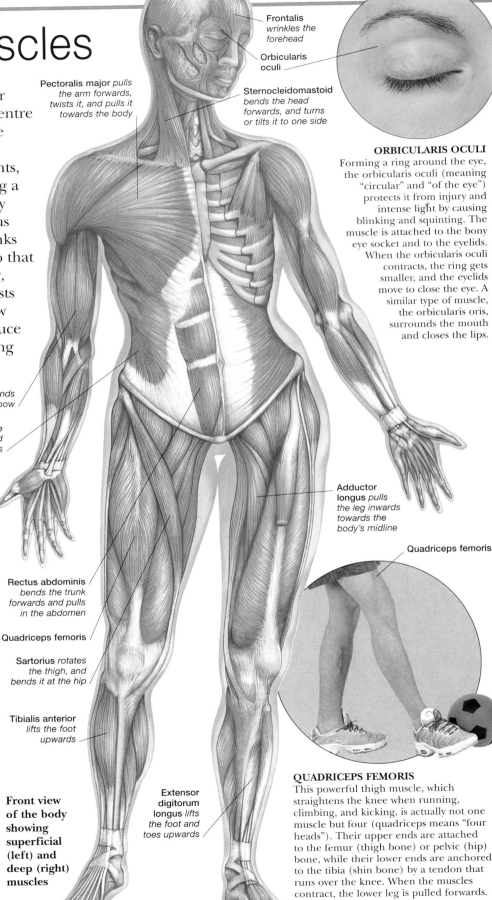

Frontalis *wrinkles the forehead*

Orbicularis oculi

Sternocleidomastoid *bends the head forwards, and turns or tilts it to one side*

Pectoralis major *pulls the arm forwards, twists it, and pulls it towards the body*

Biceps brachii *bends the arm at the elbow*

External oblique *twists the trunk and bends it sideways*

Rectus abdominis *bends the trunk forwards and pulls in the abdomen*

Quadriceps femoris

Sartorius *rotates the thigh, and bends it at the hip*

Tibialis anterior *lifts the foot upwards*

Adductor longus *pulls the leg inwards towards the body's midline*

Quadriceps femoris

Extensor digitorum longus *lifts the foot and toes upwards*

Front view of the body showing superficial (left) and deep (right) muscles

ORBICULARIS OCULI
Forming a ring around the eye, the orbicularis oculi (meaning "circular" and "of the eye") protects it from injury and intense light by causing blinking and squinting. The muscle is attached to the bony eye socket and to the eyelids. When the orbicularis oculi contracts, the ring gets smaller, and the eyelids move to close the eye. A similar type of muscle, the orbicularis oris, surrounds the mouth and closes the lips.

QUADRICEPS FEMORIS
This powerful thigh muscle, which straightens the knee when running, climbing, and kicking, is actually not one muscle but four (quadriceps means "four heads"). Their upper ends are attached to the femur (thigh bone) or pelvic (hip) bone, while their lower ends are anchored to the tibia (shin bone) by a tendon that runs over the knee. When the muscles contract, the lower leg is pulled forwards.

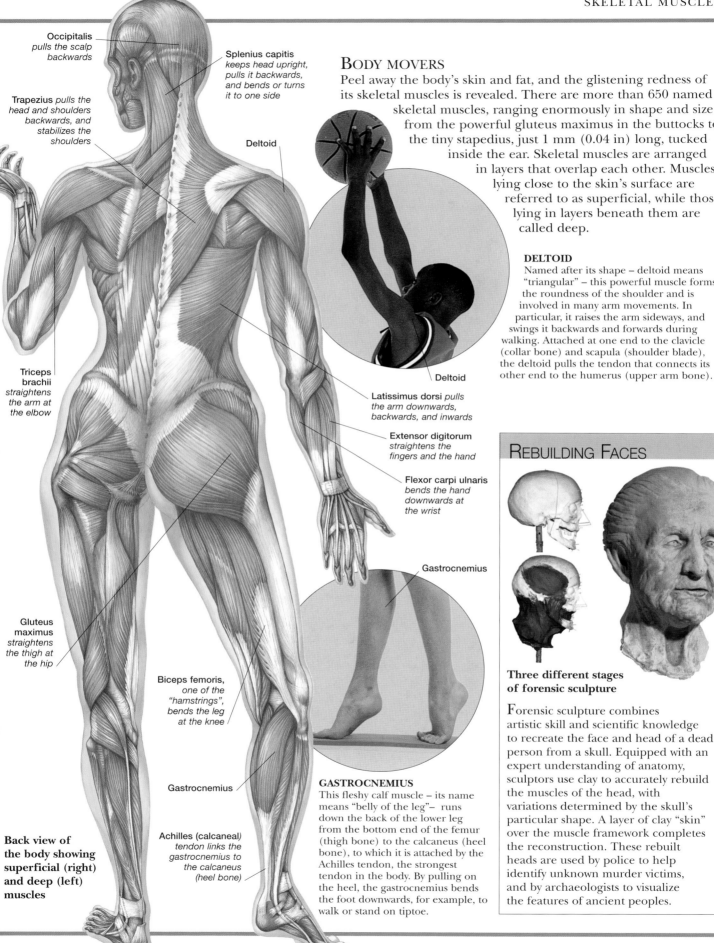

Occipitalis *pulls the scalp backwards*

Splenius capitis *keeps head upright, pulls it backwards, and bends or turns it to one side*

Trapezius *pulls the head and shoulders backwards, and stabilizes the shoulders*

Deltoid

Triceps brachii *straightens the arm at the elbow*

Latissimus dorsi *pulls the arm downwards, backwards, and inwards*

Extensor digitorum *straightens the fingers and the hand*

Flexor carpi ulnaris *bends the hand downwards at the wrist*

Gluteus maximus *straightens the thigh at the hip*

Biceps femoris, *one of the "hamstrings", bends the leg at the knee*

Gastrocnemius

Gastrocnemius

Achilles (calcaneal) *tendon links the gastrocnemius to the calcaneus (heel bone)*

Back view of the body showing superficial (right) and deep (left) muscles

BODY MOVERS

Peel away the body's skin and fat, and the glistening redness of its skeletal muscles is revealed. There are more than 650 named skeletal muscles, ranging enormously in shape and size from the powerful gluteus maximus in the buttocks to the tiny stapedius, just 1 mm (0.04 in) long, tucked inside the ear. Skeletal muscles are arranged in layers that overlap each other. Muscles lying close to the skin's surface are referred to as superficial, while those lying in layers beneath them are called deep.

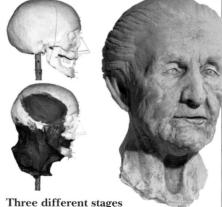

Deltoid

DELTOID

Named after its shape – deltoid means "triangular" – this powerful muscle forms the roundness of the shoulder and is involved in many arm movements. In particular, it raises the arm sideways, and swings it backwards and forwards during walking. Attached at one end to the clavicle (collar bone) and scapula (shoulder blade), the deltoid pulls the tendon that connects its other end to the humerus (upper arm bone).

REBUILDING FACES

Three different stages of forensic sculpture

Forensic sculpture combines artistic skill and scientific knowledge to recreate the face and head of a dead person from a skull. Equipped with an expert understanding of anatomy, sculptors use clay to accurately rebuild the muscles of the head, with variations determined by the skull's particular shape. A layer of clay "skin" over the muscle framework completes the reconstruction. These rebuilt heads are used by police to help identify unknown murder victims, and by archaeologists to visualize the features of ancient peoples.

GASTROCNEMIUS

This fleshy calf muscle – its name means "belly of the leg"– runs down the back of the lower leg from the bottom end of the femur (thigh bone) to the calcaneus (heel bone), to which it is attached by the Achilles tendon, the strongest tendon in the body. By pulling on the heel, the gastrocnemius bends the foot downwards, for example, to walk or stand on tiptoe.

How muscles contract

UNDERSTANDING HOW muscle fibres are constructed is key to working out how muscles contract. Each long, cylindrical muscle fibre is filled with smaller fibres, called myofibrils, which are packed with a highly ordered array of protein filaments. The arrival of a nerve message from the brain causes these filaments to interact, making the muscle fibre – and muscle – shorten. The more signals that arrive, the more a muscle contracts, until it reaches about 70 per cent of its resting length. The whole process involves the transformation of chemical energy stored in nutrients, such as glucose, into kinetic (movement) energy. In the absence of nervous stimulus, the muscle fibre relaxes.

Muscle *is an organ that contracts to move, support, or stabilize part of the body*

Muscle fibre *is one of the long, thin cells that makes up a muscle*

Fascicle *is one of the bundles of fibres that make up a muscle. It is surrounded by a connective tissue sheath, the perimysium*

Capillary *supplies muscle fibres with blood*

Sarcomere *is a section of myofibril between Z lines*

Myofibril *is one of the parallel, rod-like strands that pack the inside of a muscle fibre*

Mitochondrion

TEM OF SECTION THROUGH SKELETAL MUSCLE FIBRE

Z line

Myofibril

FROM FIBRE TO FILAMENT

Hundreds of bundled muscle fibres run in parallel along the length of a muscle. Each fibre is packed with rod-like myofibrils which contain two types of protein filament – thick (myosin) and thin (actin). These filaments are arranged in repeated patterns called sarcomeres, that give muscle fibres their striped appearance. Extending from myosin filaments are small "heads" that, in resting muscle, extend towards acting filaments.

Sarcomere

Myosin filament

Relaxed muscle fibre

Actin filament

Z line

Contracted muscle fibre

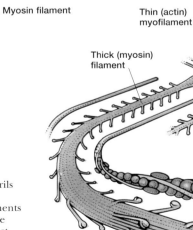

Thin (actin) myofilament

Thick (myosin) filament

Head of myosin molecule *is "charged" with ATP and interacts with actin during contraction*

ENERGY PROVIDERS

Energy-rich nutrients cannot be used directly for muscle contraction – they must first be converted into ATP (adenosine triphosphate). This substance stores energy, carries it to where it is needed, and releases it on demand. ATP is produced by aerobic respiration inside the mitochondria squeezed in between myofibrils. During respiration, glucose – delivered to muscle fibres by the blood or extracted from their glycogen store – is broken down using oxygen, releasing its energy to make ATP.

SLIDING FILAMENTS

To make contractions happen, filaments in myofibrils slide over each other. In each sarcomere, myosin filaments are located centrally while the actin filaments that surround and overlap them are attached to the Z line. When a muscle fibre is stimulated to contract, myosin heads bind to actin and, using energy supplied by ATP, repeatedly swivel towards the centre of the sarcomere, pulling actin filaments inwards and making the sarcomere shorter until the stimulus stops.

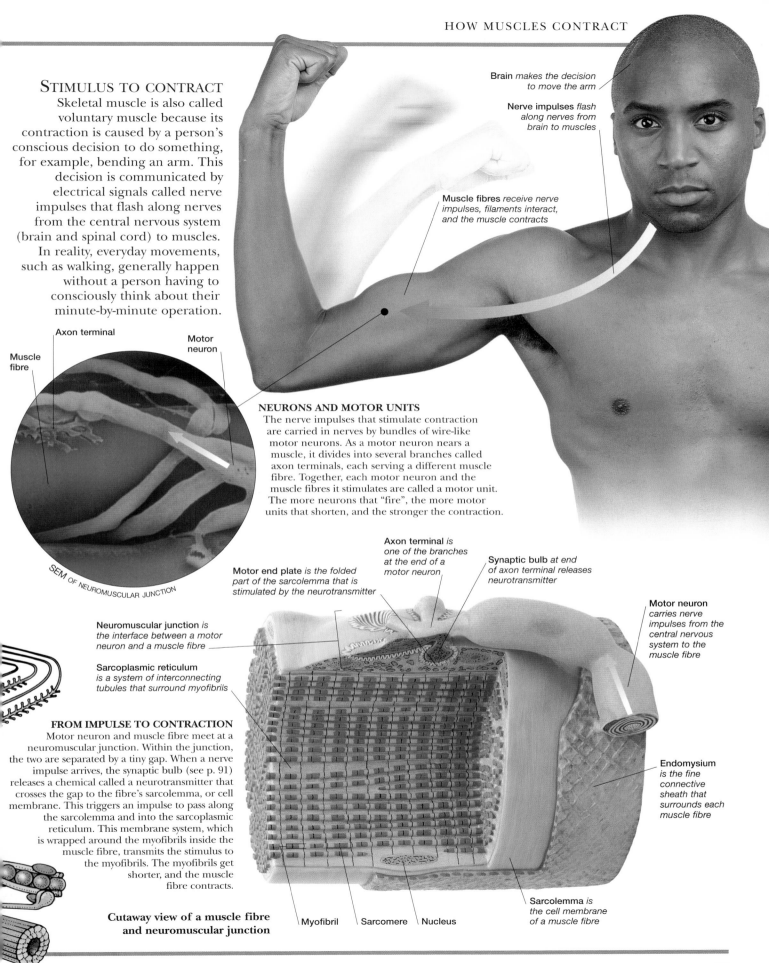

STIMULUS TO CONTRACT

Skeletal muscle is also called voluntary muscle because its contraction is caused by a person's conscious decision to do something, for example, bending an arm. This decision is communicated by electrical signals called nerve impulses that flash along nerves from the central nervous system (brain and spinal cord) to muscles. In reality, everyday movements, such as walking, generally happen without a person having to consciously think about their minute-by-minute operation.

Brain *makes the decision to move the arm*

Nerve impulses *flash along nerves from brain to muscles*

Muscle fibres *receive nerve impulses, filaments interact, and the muscle contracts*

Muscle fibre

Axon terminal

Motor neuron

SEM OF NEUROMUSCULAR JUNCTION

NEURONS AND MOTOR UNITS

The nerve impulses that stimulate contraction are carried in nerves by bundles of wire-like motor neurons. As a motor neuron nears a muscle, it divides into several branches called axon terminals, each serving a different muscle fibre. Together, each motor neuron and the muscle fibres it stimulates are called a motor unit. The more neurons that "fire", the more motor units that shorten, and the stronger the contraction.

Axon terminal *is one of the branches at the end of a motor neuron*

Synaptic bulb *at end of axon terminal releases neurotransmitter*

Motor end plate *is the folded part of the sarcolemma that is stimulated by the neurotransmitter*

Motor neuron *carries nerve impulses from the central nervous system to the muscle fibre*

Neuromuscular junction *is the interface between a motor neuron and a muscle fibre*

Sarcoplasmic reticulum *is a system of interconnecting tubules that surround myofibrils*

FROM IMPULSE TO CONTRACTION

Motor neuron and muscle fibre meet at a neuromuscular junction. Within the junction, the two are separated by a tiny gap. When a nerve impulse arrives, the synaptic bulb (see p. 91) releases a chemical called a neurotransmitter that crosses the gap to the fibre's sarcolemma, or cell membrane. This triggers an impulse to pass along the sarcolemma and into the sarcoplasmic reticulum. This membrane system, which is wrapped around the myofibrils inside the muscle fibre, transmits the stimulus to the myofibrils. The myofibrils get shorter, and the muscle fibre contracts.

Endomysium *is the fine connective sheath that surrounds each muscle fibre*

Sarcolemma *is the cell membrane of a muscle fibre*

Cutaway view of a muscle fibre and neuromuscular junction

Myofibril | Sarcomere | Nucleus

ANATOMY AND ART

FOR THE PAST 500 YEARS, the science of anatomy – the study of the structure of the human body – and the art of anatomical illustration have developed hand in hand. Artists and doctors have combined efforts to produce visual representations of how muscles, bones, nerves, and other structures form the fabric of the body. The main purpose of such images has been to aid the teaching of anatomy to medical students by describing and explaining the confusing array of tissues and organs they come across inside bodies, both living and dead. Yet the sheer beauty of many of these illustrations and sculptures means that they have wider appeal.

SIMPLE IMAGES
This 15th-century image of the female body makes no pretence at accuracy. It was used to show wounds, diseases, and the influence of the zodiac on body parts.

EARLY BELIEFS

Until the Renaissance – the "rebirth" of arts and sciences between the 14th and 16th centuries – anatomical art was flat and schematic, owing less to reality and more to myth and astrology. Knowledge about the body was still firmly rooted in the ancient and often inaccurate teachings of the Greek physician Claudius Galen (c. ad 130–200). Only during the Renaissance, when dissection (cutting up) of bodies was finally allowed, did some people start to question Galen's legacy.

CLOSE OBSERVATION
Both scientist and artist, Leonardo da Vinci used his great skills as an observer, draughtsman, and anatomist to produce over a thousand drawings of the human body, based on his own dissections.

ANATOMICAL ACCURACY
Above is an accurate representation of the muscular system, taken from Vesalius's De Humani Corporis Fabrica, the first great book of anatomy to be published. In the style of the book, the figure is standing in a life-like pose and is pictured against a landscape.

LEONARDO DA VINCI

One of those dissectors was Italian artist and scientist Leonardo da Vinci (1452–1519). Cutting up bodies by candlelight, Leonardo meticulously recorded his observations in the form of detailed drawings that showed both depth and realism. But he never took advantage of the recent inventions of engraving and printing. His drawings remained unseen until his notebooks were rediscovered in the 19th century.

ANDREAS VESALIUS

The breakthrough in anatomical illustration came in 1543 with the publication of *De Humani Corporis Fabrica* (On the Structure of the Human Body) by Flemish doctor Andreas Vesalius (1514–64), who was based in

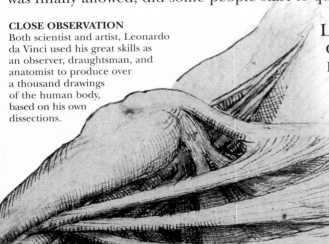

Superficial muscles, drawn by Leonardo da Vinci

Padua, Italy. Over 300 illustrations by Jan Stefan van Kalkar, a favourite student of the Venetian artist Titian, used artistic form to show the detail of human anatomy exposed by Vesalius's careful and accurate dissections. By questioning and correcting Galen's works, Vesalius established the science of anatomy based on observation and realism, and initiated a new era in medicine.

THE SEARCH FOR REALISM

After Vesalius, artists sought to depict the dissected body realistically, as if it were still alive. As well as sumptuous drawings, there was a fashion for wax models. These models gave medical students a three-dimensional view of the body but, unlike a dissected body, the wax sculpture did not rot and smell, and it could be coloured to show clearly an individual muscle's nerves and blood vessels.

ANATOMY MANUALS

The 19th century brought with it printing and colouring techniques that could be used to mass-produce well-illustrated manuals of anatomy compiled by eminent anatomists and distinguished artists. Some, like the *Traité Complet de l'Anatomie d'Homme* (Complete Treatise of the Anatomy of Man) (1831–54) by Jean-Baptiste Bougéry (doctor) and Nicolas Jacob (artist), maintained the tradition of showing life-like bodies.

MODELLED IN WAX
This 19th-century wax model, made in Italy, shows in incredible detail and vibrant colours the muscles, bones, blood vessels, and nerves of the upper limb.

19th-century wax model made at the Museum "la Specola" in Florence, Italy

MODERN METHODS

Some of today's anatomy manuals still use the beauty of artistic form to modify a raw dissection in order to reveal anatomy and function. But the medical illustrator now has other tools at his or her disposal. Computers can be used to create two- and three-dimensional images of the body, integrating input from drawings, photographs, or from imaging techniques such as CT and MRI scans (see pp. 30–1).

COMPUTER GENERATION
Shown in full on p. 167, this computer-generated image of the human respiratory (breathing) system was built up layer by layer using image-making software, and then overlaid on the photograph of a real person.

TEXTBOOK ILLUSTRATIONS
This detailed illustration of a dissected armpit from Bougéry and Jacob's 19th-century anatomical work was printed by lithography, a technique invented in 1798 that made it possible to produce inexpensive, well-illustrated medical textbooks.

Movement and posture

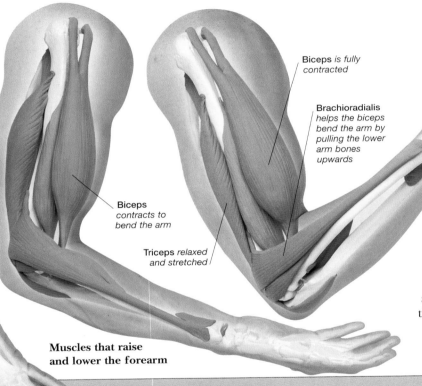

IF ITS MUSCLES WERE suddenly inactivated, the body would not only be immobilized but would also collapse. As well as moving the body, skeletal muscles hold it upright and maintain posture. To perform both roles, muscles pull the bones, to which they are attached by tendons, across a joint. When a muscle contracts, one of the bones to which it is attached – the insertion – moves, while the other attachment point – the origin – remains fixed. Since they can only pull and not push, muscles work in antagonistic pairs to produce opposing movements. Flexor muscles in the forearm, for example, bend fingers, while their antagonists, the extensor muscles, straighten them. Muscles that work together to produce the same movement are called synergists.

Triceps *contracts to straighten the arm*

Biceps *is fully contracted*

Brachioradialis *helps the biceps bend the arm by pulling the lower arm bones upwards*

Biceps *contracts to bend the arm*

Triceps *relaxed and stretched*

Muscles that raise and lower the forearm

OPPOSING MUSCLES

When a muscle contracts it shortens, pulling its insertion towards its origin. To produce movement in the opposite direction, there must be a separate antagonistic, or opposing, muscle. In the arm, for example, the biceps muscle pulls the forearm upwards towards the shoulder to bend the arm. Its antagonist, the triceps, pulls the forearm downwards to straighten the arm. The brachioradialis acts as a synergist to the biceps, helping it to bend the arm.

MUSCLES, BONES, AND LEVERS

Muscles and bones interact to move the body using lever systems. A lever is a bar that moves on a fixed point, the fulcrum, when a force is applied to one part of it to move a weight on another. In the body, bones are levers, a joint is a fulcrum, and muscle contraction provides the force to move a body part (weight). Levers fall into three classes according to the relative positions of the force, weight, and fulcrum.

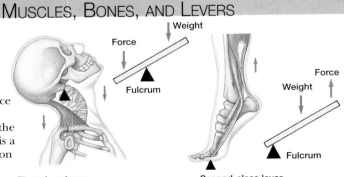

First-class lever
The fulcrum lies between force and weight, like a see-saw. Neck muscles pulling the back of the skull to tilt the head backwards produce a similiar action.

Second-class lever
The weight lies between the force and the fulcrum, like a wheelbarrow – for example, raising of the heel (and body weight) by the calf muscles.

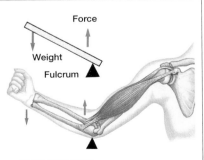

Third-class lever
The most common type of lever in the body involves force applied between fulcrum and weight, like tweezers – for example, bending the elbow.

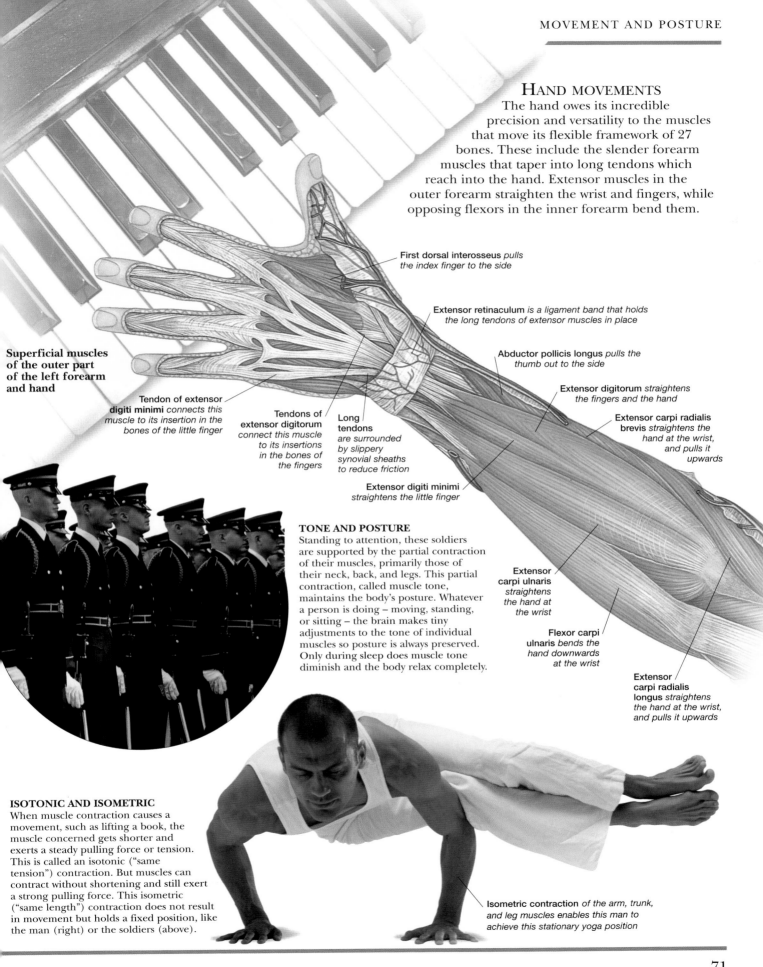

HAND MOVEMENTS

The hand owes its incredible precision and versatility to the muscles that move its flexible framework of 27 bones. These include the slender forearm muscles that taper into long tendons which reach into the hand. Extensor muscles in the outer forearm straighten the wrist and fingers, while opposing flexors in the inner forearm bend them.

First dorsal interosseus *pulls the index finger to the side*

Extensor retinaculum *is a ligament band that holds the long tendons of extensor muscles in place*

Abductor pollicis longus *pulls the thumb out to the side*

Extensor digitorum *straightens the fingers and the hand*

Extensor carpi radialis brevis *straightens the hand at the wrist, and pulls it upwards*

Superficial muscles of the outer part of the left forearm and hand

Tendon of extensor digiti minimi *connects this muscle to its insertion in the bones of the little finger*

Tendons of extensor digitorum *connect this muscle to its insertions in the bones of the fingers*

Long tendons *are surrounded by slippery synovial sheaths to reduce friction*

Extensor digiti minimi *straightens the little finger*

Extensor carpi ulnaris *straightens the hand at the wrist*

Flexor carpi ulnaris *bends the hand downwards at the wrist*

Extensor carpi radialis longus *straightens the hand at the wrist, and pulls it upwards*

TONE AND POSTURE

Standing to attention, these soldiers are supported by the partial contraction of their muscles, primarily those of their neck, back, and legs. This partial contraction, called muscle tone, maintains the body's posture. Whatever a person is doing – moving, standing, or sitting – the brain makes tiny adjustments to the tone of individual muscles so posture is always preserved. Only during sleep does muscle tone diminish and the body relax completely.

ISOTONIC AND ISOMETRIC

When muscle contraction causes a movement, such as lifting a book, the muscle concerned gets shorter and exerts a steady pulling force or tension. This is called an isotonic ("same tension") contraction. But muscles can contract without shortening and still exert a strong pulling force. This isometric ("same length") contraction does not result in movement but holds a fixed position, like the man (right) or the soldiers (above).

Isometric contraction *of the arm, trunk, and leg muscles enables this man to achieve this stationary yoga position*

71

Muscles and exercise

THE BODY IS capable of responding to all manner of changes, both inside and outside itself. One of the most obvious responses is to increased activity or exercise. The heart, lungs, and muscles undergo changes – all carefully regulated by the brain – that ensure that muscle fibres obtain sufficient energy to contract more rapidly and more strongly. The ability of the body to react in this way depends on its fitness. A fit body is one that can carry out everyday activities, such as running for a bus or climbing stairs, without breathlessness or excessive tiredness. Fitness has three elements – stamina, strength, and flexibility – all of which can be increased by regular exercise. Some exercises focus on just one of these fitness elements, while others, such as swimming, improve all three.

Muscles *work harder during exercise and demand more energy*

Fibres in the arm's muscles increase in size, which improves strength

Weightlifting is an anaerobic exercise that increases strength

STRENGTH

The amount of force muscles exert when performing an action such as lifting or jumping, and their ability to hold the body upright without tiring, are determined by their strength. Exercises such as weightlifting improve strength by increasing the size of muscle fibres. These exercises are anaerobic. That is, they are intensive exercises that last for a very short time, and use energy released without the need for oxygen. They do not improve stamina.

STAMINA

Also called cardiovascular fitness or endurance, stamina reflects the ability of the heart and blood vessels to deliver oxygen and nutrients to the body's tissues, including muscles. Stamina is enhanced by aerobic exercises, such as running, brisk walking, or dancing, which use oxygen to release energy from "fuels" such as glucose. Performed for 20 minutes or more, at least three times a week, and demanding enough to produce sweating and slight breathlessness, aerobic activities improve stamina by increasing the strength and efficiency of the heart.

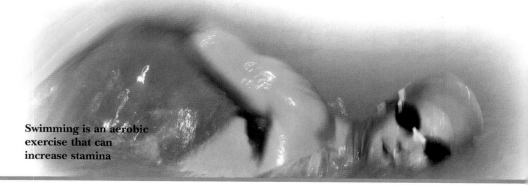

Swimming is an aerobic exercise that can increase stamina

ON THE MOVE

When exercise begins, several changes occur automatically to meet the demands for extra oxygen and nutrients needed by skeletal muscles that are working harder than normal. Blood flow to muscle fibres is dramatically increased during vigorous exercise from 1 to 12 litres (1.8 to 21.1 pints) per minute. This is achieved by widening blood vessels that serve muscle fibres, and by making the heart pump faster and more strongly so that blood flow rate increases from 5 to 20 litres (8.8 to 35.2 pints) per minute. At the same time, the volume of air taken in by the lungs goes from 6 up to 100 litres (0.2 to 3.5 cu ft) per minute by increasing the rate and depth of breathing.

"Scorpion" yoga position

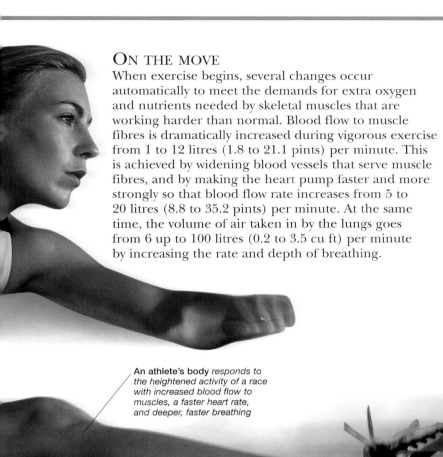

An athlete's body *responds to the heightened activity of a race with increased blood flow to muscles, a faster heart rate, and deeper, faster breathing*

Mitochondrion

Myofibril

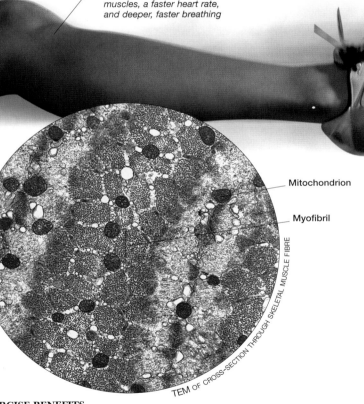

TEM OF CROSS-SECTION THROUGH SKELETAL MUSCLE FIBRE

FLEXIBILITY

The body's flexibility is measured by the ability of its joints to move freely, and without any discomfort. Activities that improve flexibility – including yoga and gymnastics – involve stretching and holding exercises, and ensure that muscles are supple, and that ligaments and tendons remain in good working order. Stretching the body just after exercise, while muscles are still "warmed up", also helps avoid muscle pain and stiffness.

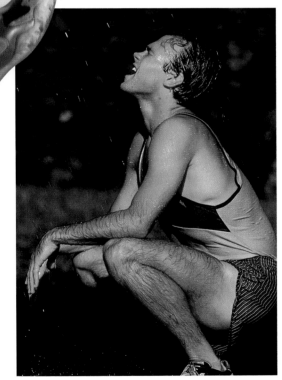

EXERCISE BENEFITS

Regular exercise which combines elements that increase stamina, strength, and flexibility has numerous benefits. The heart pumps, and the lungs take in oxygen, more efficiently. Capillaries supplying muscle fibres with nutrients and oxygen increase in number. More mitochondria (organelles that release energy) and glycogen granules (energy stores) appear in muscle fibres, and there is more myoglobin (the substance that carries oxygen). Muscle fibres increase in size and work more efficiently, giving muscles greater strength and resistance to tiredness.

OXYGEN DEBT

An athlete pants after a race because his body needs to get extra oxygen to its cells – over and above their resting oxygen consumption – in order to "pay off" an "oxygen debt". This "debt" is generated during hard exercise by anaerobic respiration, when muscle fibres obtain energy without using oxygen. The waste product of this process – lactic acid – must be disposed of by aerobic respiration, a process that requires extra oxygen, before muscles can work normally again.

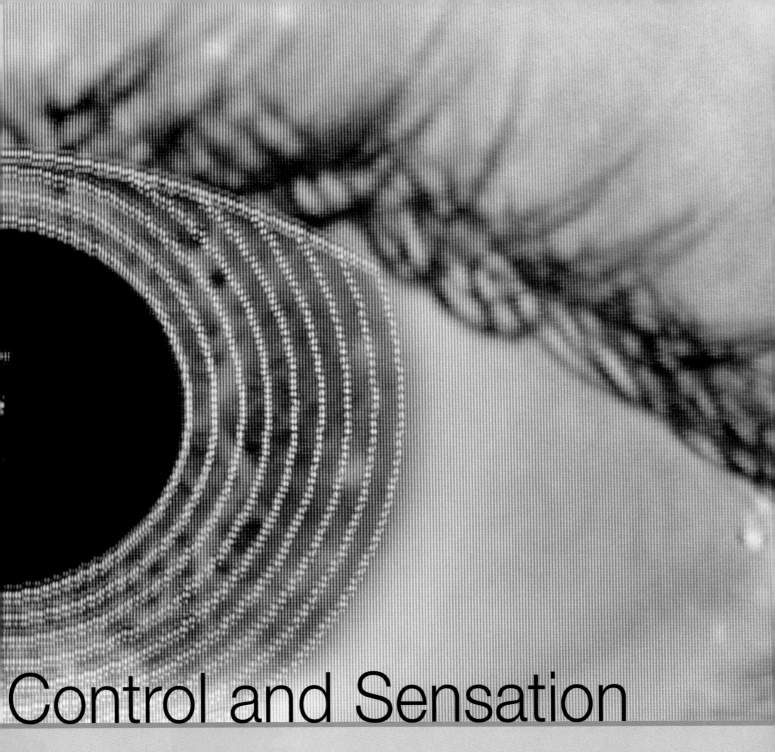

Control and Sensation

THE HUMAN BRAIN is a remarkable organ. It can simultaneously order leg muscles to run and the heart to beat faster. It gives humans their creativity, memory, intelligence, emotions, and personality. It turns messages received from sensors into the sensations that allow humans to see, hear, taste, smell, and touch. Together with its nerves and sensors, the brain forms the nervous system. A second, linked control system – the endocrine system – regulates growth, reproduction and some other processes.

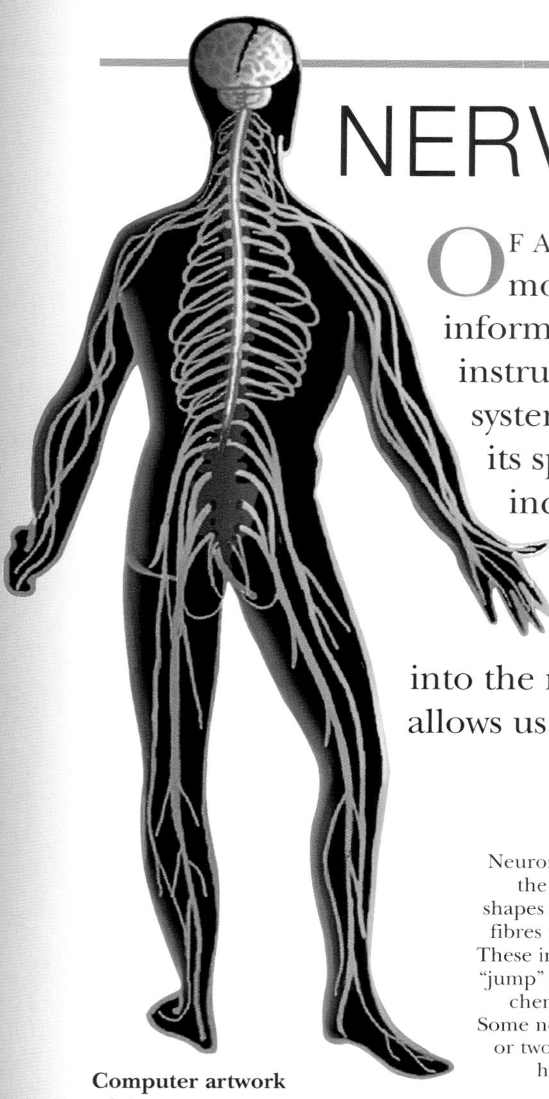

NERVOUS system

O F ALL THE HUMAN body's systems, the nervous system is the most complex. It is on duty every second of every day, gathering information about the body and its surroundings, and issuing instructions that make the body react. Together with the endocrine system (see p. 118), it controls everything that the body does, and its speed and processing power mean that it can cope with an incredible range of tasks at the same time. It works through specialized cells called neurons, which carry signals in the form of tiny bursts of electricity. Some neurons carry signals to, or from, particular parts of the body, but most are packed into the nervous system's headquarters – the brain. This living computer allows us to think and to remember, and makes us who we are.

**Computer artwork
of the nervous system**

NEURONS

Neurons, or nerve cells, are the basic units of the nervous system. They come in many shapes and sizes, but they all have slender fibres that can carry electrical impulses. These impulses flash along the cell, and "jump" from one neuron to another at chemical junctions called synapses. Some neurons have synapses with one or two neurons, but others can have hundreds of these connections.

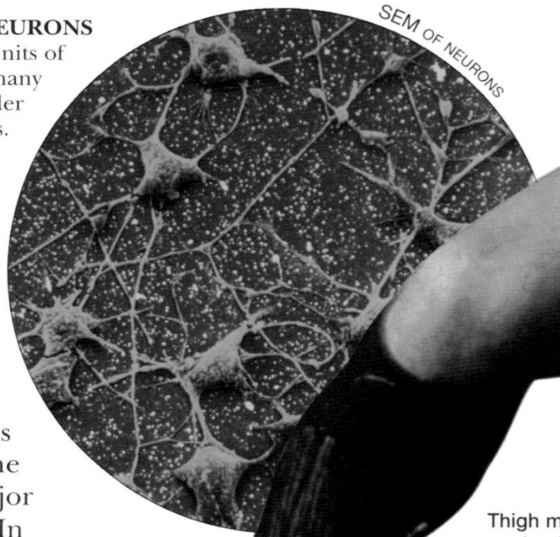

SEM OF NEURONS

COMMUNICATION NETWORK

The nervous system reaches almost every part of the body, from muscles and sense organs to the inside of teeth and bones. Nerves are the system's main communication cables, fanning out from the spinal cord, and also from the brain. There are more than 80 major nerves, and each one may contain more than a million neurons. In the above diagram, the central nervous system, comprising the brain and spinal cord, is coloured blue-white; the spine, shown around the spinal cord, is coloured magenta; and the spinal nerves are orange.

Thigh muscle
contracts to straighten leg when stimulated by nerve impulses from the brain carried by motor neurons

Pressure receptors
in skin detect the force with which the foot pushes down on the ground

NERVOUS SYSTEM FUNCTIONS

Sensory	*Senses changes (stimuli) inside and outside the body, in conjunction with receptors or sense organs. The changes include a wide range of physical factors, such as light, pressure, or the concentration of dissolved chemicals.*	Integrative	*Analyses sensory information, and makes decisions on appropriate responses. Triggered or modified by information that is stored in and retrieved from memory.*
		Motor	*Triggers responses by muscles or glands. The nervous system can either stimulate muscles and glands into action, or it can inhibit them.*

discovered that he also had brain damage on the left side, but further back. He concluded that this region – called Wernicke's area – dealt with choosing the right words to speak.

INVASIVE TECHNIQUES

Some researchers took a more direct approach. In 1870, when German army surgeon Eduard Hitzig (1838–1907) was treating wounded soldiers whose brains were exposed, he applied electrical currents to different parts of their brains and noted what happened. In the 1950s, Canadian brain surgeon Wilder Penfield (1891–1976) carried out similar, but more sophisticated, experiments. During surgery, when a patient's brain was exposed – but he or she was conscious – Penfield electrically stimulated different parts of the brain and carefully recorded which regions produced sensations, caused movements, or stored memories.

HAND CONTROL
These two MEG scans of the left side of the brain show brain–hand control in real time. On the left, milliseconds (msecs) before the person moves their right index finger, neurons in the motor cortex "light up" as they send instructions to the finger-moving muscles. On the right, just 40 msecs later, the sensory cortex "lights up" as it receives a message from the muscles that the finger is moving.

SCANNING THE BRAIN

Today's brain mappers use non-invasive methods to investigate brain function. PET and MRI scans (see pp. 30–31) produce indirect images of activity in living brains by measuring oxygen uptake or blood flow in brain regions that are active. MEG (magnetoencephalography) scans measure brain activity directly and in real time – as it happens – by monitoring the electrical activity of brain cells themselves. What all these techniques have shown is that the brain is made up of interactive, not isolated, units.

VICTIMS OF WAR
War inevitably produces head wounds, as shown by these soldiers from the First World War. Some wounded soldiers became subjects for researchers who were interested in using their exposed brains to find out which area controlled which action.

Soldiers wounded in the Somme Offensive, 1916

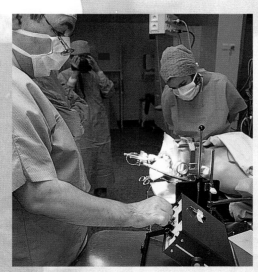

BRAIN PROBE
Surgeons use a frame fixed to a patient's head to provide support and guidance for probes that will go into specific parts of the brain through holes drilled in the skull. These probes are used to remove diseased tissue, but have also been used to find out more about brain function.

Cerebral activities

Humans are not alone in having complex brains, but they are the only living things that have some idea about how their brains actually work. This ability – and countless others as well – comes about because our cerebral hemispheres are exceptionally well developed. Together, the two hemispheres give the human brain its amazing ability to process and store information, and they also initiate every deliberate movement that the body makes. These different tasks are carried out by grey matter, which forms the surface of the brain (cerebral cortex) and "islands" deep inside it. Each region of grey matter carries out a particular function, and different regions work together to perform complex actions, such as reading and understanding this page. Together, they also generate consciousness – our sense of self-awareness.

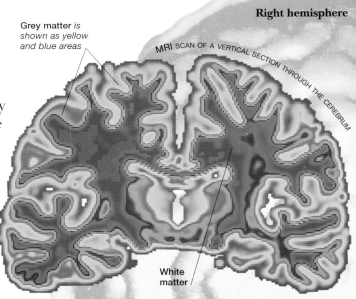

Right hemisphere

Grey matter *is shown as yellow and blue areas*

MRI SCAN OF A VERTICAL SECTION THROUGH THE CEREBRUM

White matter

GREY MATTER
This vertical slice through the brain shows grey matter spread over the surface or cortex of the cerebral hemispheres, with white matter beneath it. Grey matter consists of neuron cell bodies, while white matter consists of axons – the brain's equivalent of "wiring". These axons connect neurons with each other, and they also carry signals to and from other parts of the body.

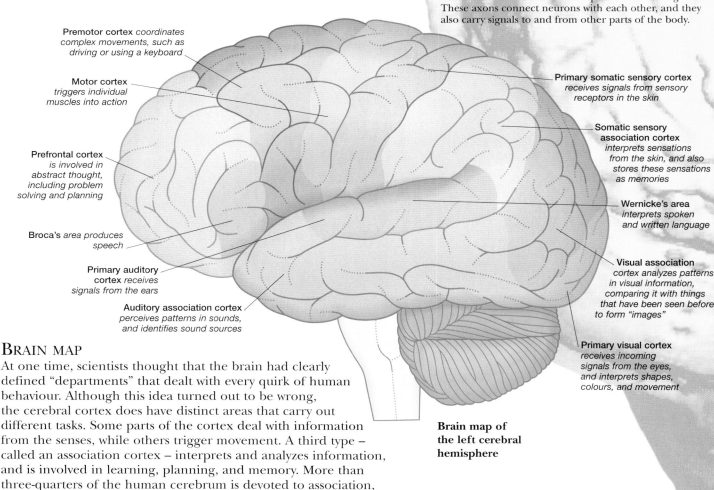

Premotor cortex *coordinates complex movements, such as driving or using a keyboard*

Motor cortex *triggers individual muscles into action*

Prefrontal cortex *is involved in abstract thought, including problem solving and planning*

Broca's *area produces speech*

Primary auditory cortex *receives signals from the ears*

Auditory association cortex *perceives patterns in sounds, and identifies sound sources*

Primary somatic sensory cortex *receives signals from sensory receptors in the skin*

Somatic sensory association cortex *interprets sensations from the skin, and also stores these sensations as memories*

Wernicke's area *interprets spoken and written language*

Visual association *cortex analyzes patterns in visual information, comparing it with things that have been seen before to form "images"*

Primary visual cortex *receives incoming signals from the eyes, and interprets shapes, colours, and movement*

Brain map of the left cerebral hemisphere

Brain map

At one time, scientists thought that the brain had clearly defined "departments" that dealt with every quirk of human behaviour. Although this idea turned out to be wrong, the cerebral cortex does have distinct areas that carry out different tasks. Some parts of the cortex deal with information from the senses, while others trigger movement. A third type – called an association cortex – interprets and analyzes information, and is involved in learning, planning, and memory. More than three-quarters of the human cerebrum is devoted to association, compared with less than a quarter in many other mammals.

Left hemisphere

CT SCAN OF TOP OF BRAIN

LEFT AND RIGHT

The cerebrum is divided into left and right halves, or hemispheres, connected by a "bridge" of nerve fibres called the corpus callosum. The left hemisphere controls the body's right side, and the right hemisphere controls the left side. Usually the left hemisphere is dominant, which is why most people are right-handed. The left hemisphere is in charge of written and spoken language, numbers, and problem solving, among other things, while the right hemisphere deals with appreciation of art and music, and recognizing faces.

REACTION TIMES

In tennis and other fast-moving sports, rapid reactions are essential. For example, a tennis player may have less than half a second in which to judge the most appropriate movement to return a serve. This is not enough time to make a conscious decision about exactly how to move; the player's brain works subconsciously, triggering the right movements before the player becomes aware of having decided what to do.

Thalamus *relays incoming information to the cerebal cortex*

Motor neuron cell body

Motor cortex

Caudate nucleus

Putamen

Basal ganglia *plan, initiate, and monitor complex movement (links to other parts of the brain not shown)*

Globus pallidus

White matter *contains mainly nerve fibres*

Grey matter *contains neuron cell bodies and dendrites*

Cerebellum *co-ordinates body posture and muscular movements*

Brain stem *relays messages between spinal cord and brain*

Spinal cord

PRECISION MOVEMENT

Even a simple action, such as picking up a small object, involves several different parts of the brain. Movement is triggered by the motor cortex, which sends signals to muscles in the arm. Once it is underway, sensory signals travel back to the motor cortex, via the cerebellum and thalamus, where they modify the movement, making muscles contract by exactly the right amount. This kind of feedback control allows us to pick up delicate things without breaking them.

Skeletal muscles *receive instructions from the cortex and cerebellum, and move the hand*

Sensory receptor *in muscle detects tension*

Voluntary movements *are continually corrected by messages from the cerebellum*

Signal sent from:

Motor cortex to muscle

Muscle's sensory receptor to cerebellum

Cerebellum via thalamus to cortex

Cerebellum via spinal cord to muscle

93

Memory and emotions

DESPITE THE HUGE SIZE of the human population, no two people behave in exactly the same way. One reason for this is that we learn from experience, and another is that emotions affect the way we respond. Memory and emotions both involve the limbic system, as well as other areas of the brain. Emotions are instant responses, but memory works in stages. Things that we experience pass first into the sensory memory, which holds them for a few seconds, and then into the short-term memory, which can store them for minutes or hours. Finally, they enter the long-term memory, which can store information for an entire lifetime. To prevent the brain being overwhelmed by data, only about one per cent of what we experience actually reaches this final stage.

Limbic system (blue) shown within the cerebral hemispheres

Cingulate gyrus *is the part of the limbic cortex that modifies behaviour and emotions*

Thalamus

Fornix *is a pathway of nerve fibres that links different parts of the limbic system together*

Cerebrum

Olfactory bulb

Amygdala

Hippocampus

Parahippocampal gyrus *is the part of the limbic cortex that modifies emotions such as rage and fright*

THE LIMBIC SYSTEM

Without the limbic system, life would be more hazardous, and also much less interesting. The reason is that this part of the brain guards us against everyday hazards, and also controls sexual instincts, pleasure, and pain. The limbic system is deep within the brain, and it consists of a curve of structures that stretch around the brain stem. Among them are the hippocampus, which plays a part in establishing memories, and the amygdala, which triggers an inbuilt fear of danger, such as falling or being attacked. The olfactory bulbs are also part of the limbic system, which is why smells can trigger off memories and emotions.

Playing the violin depends on learned skills stored in procedural memory

Sensory memory
Retains input from, say, sights or sounds for a few seconds.

Short-term memory
Receives sensory input, and may retain and interpret it.

Consolidation
Transfer from short-term to long-term memory by the hippocampus.

Long-term memory
Provides both memory storage and retrieval, and has three basic forms.

Information forgotten
Most information is stored only briefly, then lost.

Procedural memory
Involves skills learned through practice, such as riding a bicycle.

Semantic memory
Deals with words, language, facts, and their meanings.

Episodic memory
Records specific events and experiences, such as a holiday.

MEMORY

Sensory memory makes a person aware of their surroundings from moment to moment. This input is briefly stored in short-term memory, where thoughts, words, and emotions are analysed. Most short-term memories are quickly lost, but "significant" ones are shunted to the hippocampus. Here, they are replayed repeatedly, over weeks or years, to the cerebral cortex. Each repetition etches the event, experience, or skill ever deeper until it is lodged in long-term memory.

Inner surface *of right cerebral hemisphere*

CT SCAN OF SECTION THROUGH BRAIN

Hypothalamus

HYPOTHALAMUS

Positioned near the base of the brain, the hypothalamus plays a key part in emotional responses, such as anger or fear. It does this by controlling the autonomic nervous system, and also by releasing hormones into the nearby pituitary gland. Linked to the limbic system (and sometimes described as part of it), the hypothalamus also acts as the body's thermostat, its energy and water manager, and its clock.

Laughter reflects a state of happiness

EMOTIONS

Anger, amusement, sadness, and disappointment are all examples of emotions. These deep-seated feelings – produced by the interaction of the limbic system and the cerebral cortex – help to make us human, and seem to have no direct equivalent in the animal world. Although emotions are mental states, they often express themselves through body language. Crying and laughter are two common examples.

PHOBIAS

Fear is a vital part of everyday life, because it keeps us on guard against potentially dangerous situations. It usually matches the potential threat, which means that we are most fearful of things that can do us the most harm. But some people develop an irrational fear of things that are largely harmless – a condition known as a phobia. Phobias can be triggered by all kinds of objects and situations, from spiders and mice to open spaces and air travel. In severe cases, medical treatment is needed to help sufferers lead a normal life.

Arachnophobia is the fear of spiders

Arms *are used to help balance the body*

Brain *remembers "errors" such as falling over, so they can be improved on next time*

Child *is ready to progress from standing on two feet to walking*

LEARNING

Humans are born with some instincts, but most of our behaviour – and all of our knowledge – is learned. In infants, learning takes place by trial and error, which is how a baby discovers how to walk. But as the brain develops, insight and imagination become increasingly important. Learning creates electrical pathways between neurons in the brain. Repetition reinforces these pathways, making information "stick" in the memory.

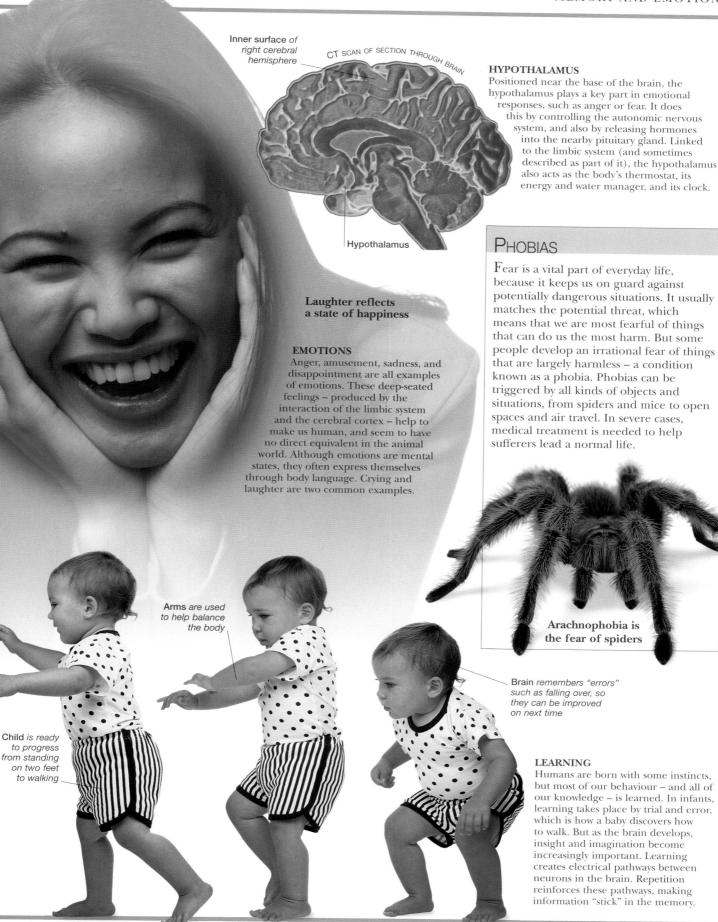

Consciousness and sleep

EVERY DAY, THE BRAIN switches between two quite different states – being awake and fully conscious, and being asleep. When the brain is awake, it is aware of its surroundings and is able to think in a purposeful way, so that it can meet the demands of daily life. During sleep, its activity is reduced, and its thought patterns are largely disconnected from the outside world. Sleep allows the body to rest, but just as importantly, it gives the brain time to sort and store the information it has accumulated during the day. Consciousness and sleep are both triggered and maintained by neurotransmitters. These chemicals are made and released by the reticular activating system – a regulator located in the brain stem at the base of the brain.

Alpha waves *occur when someone is awake but resting, or in light sleep*

Beta waves *occur when someone is awake and mentally alert*

EEG OF BRAIN WAVE PATTERNS

Delta waves *occur during deep sleep*

BRAIN WAVES

The invention of the electroencephalograph in the 1920s revealed the existence of brain waves – patterns of electrical activity created by the combined output of millions of neurons. Alpha waves are produced when the brain is awake but relaxed, while beta waves occur during times of intense activity. Widely spaced delta waves occur during deep sleep, when the brain is at its least active. The reading produced by the instrument is caused an electroencephalogram (EEG).

STAYING ALERT

When a person is awake, their brain deals with a torrent of sensory information from inside and outside the body. To do this, its association neurons have to work with maximum efficiency. This level of activity is triggered by neurons in the reticular activating system (RAS). These reach throughout the cerebral cortex – the "thinking" part of the brain – where they release acetylcholine, a neurotransmitter that aids alertness and concentration. The brain uses many other neurotransmitters as well. One of them, called serotonin, triggers sleep and also affects mood. Another, called dopamine, helps to regulate movements. People suffering from Parkinson's disease have low levels of dopamine, causing weakness and muscle tremors.

Cell bodies of neurons *in RAS produce the neurotransmitter acetylcholine*

Acetylcholine *travels along axons which reach all parts of the cerebral cortex – the folded outer layer of the brain*

Radiating signals

Visual impulses *from the eye stimulate the RAS*

RAS *in the brain stem*

Auditory impulses *from the ear stimulate the RAS*

Nerve impulses *travelling up the spinal cord stimulate the RAS*

Section of the brain showing the reticular activating system (RAS) and its nerve fibre pathways

SCIENCE IN MIND

In the 19th century, large mental hospitals were established which enabled doctors to observe patients and catalogue a wide range of disorders. The work of Broca and Wernicke (see pp. 90–1) proved that abnormal brain structure could alter behaviour, indicating that the mind really was part of the brain. In Paris, physician Jean-Martin Charcot (1825–93) drew together all that had been learned about mental illness and the brain to create a scientific approach to understanding the mind.

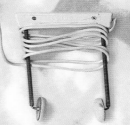

SHOCK TREATMENT

An electroconvulsive therapy (ECT) machine from 1950 is attached to a headset complete with paired electrodes. These padded electrodes were placed on the side of the head in order to deliver an electrical pulse – or shock – to the brain.

PRESSURE ON THE BRAIN

This MRI scan of a vertical section through the brain (orange) shows an abnormal growth, or tumour (blue), in the right hemisphere. Tumours like these may cause changes in behaviour and sensation, but can, in some cases, be removed by surgery.

PROBING THE UNCONSCIOUS

In 1885, an Austrian doctor called Sigmund Freud visited Charcot in Paris. On returning to Vienna, Freud evolved a method by which some patients could discuss their problems by free association – talking openly about their feelings and emotions to a person, called the analyst. Freud believed that problems were caused by subconscious fears, worries, and conflicts, often developed in childhood. His technique, called psychoanalysis, raises problems to a conscious level – by talking about them – so that the patient, with the help of the analyst, can deal with them.

THE MODERN APPROACH

But by the middle of the 20th century, there were still few treatments for mental illnesses, because there was little idea of their causes. Electroconvulsive therapy (ECT), for example, was used to treat severe depression, but its mode of action was unknown. But since then there have been considerable advances. Drugs have been developed that treat specific mental illnesses. Research has shown that in many cases these drugs work because they correct imbalances of neurotransmitters (see pp. 80–1) in the brain. Modern scanning techniques, such as MRI and PET scans, enable doctors and researchers to look for abnormal brain structures and to watch brain activity in action. Psychotherapy allows people to talk through and address their problems.

FOUNDER OF ANALYSIS

Austrian doctor Sigmund Freud (1856–1939), seen here at his desk, developed the earliest form of psychotherapy – psychoanalysis – to treat mental health problems. Freud left Austria when the Nazis took over in 1938, and moved to London.

PSYCHOTHERAPY

A person with a mental or emotional problem seeks help from a psychotherapist. The person is encouraged to talk about their symptoms and problems, while the therapist is trained to listen to and evaluate what has been said, and help the person to understand themselves more.

Touch

OF ALL THE SENSES, touch is the most direct because it tells us about the state of our own bodies. It allows us to perceive an amazing range of physical sensations, from the pain of a sudden blow to the movement of individual hairs when they are ruffled by the breeze. It is also responsible for registering heat and cold, and for some of the sensory signals that tell us about our posture – in other words, how our bodies are arranged. Unlike the special senses – such as hearing and vision – touch is a general sense, meaning that it works through receptors that are scattered all over the body. There are several kinds of receptors, and each one responds to a different type of stimulus, contributing to the sense of touch as a whole.

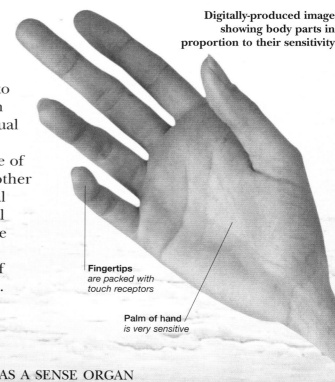

Digitally-produced image showing body parts in proportion to their sensitivity

Fingertips *are packed with touch receptors*

Palm of hand *is very sensitive*

Dermis

Meissner's corpuscle *senses light touch*

Epidermis

Merkel's disc *senses light touch and pressure, helping to pinpoint its source*

Free nerve endings *close to the skin's surface detect heat, cold, and pain*

Pacinian corpuscle *responds to vibration and firm pressure deep in the skin*

Ruffini's corpuscle *responds to continuous touch on the skin and deeper tissue*

Fat

SKIN AS A SENSE ORGAN

The skin contains more sensory receptors than any other organ in the body. Most of them are mechanoreceptors, which trigger nerve impulses when they are physically pulled or pressed. The structure and position of these receptors affects what they detect. Merkel's discs, for example, are small and are close to the surface of the skin – the ideal combination for detecting light touch and pinpointing its exact source. Pacinian corpuscles, on the other hand, are larger and deeper, and respond to firmer pressure and to being stretched. The skin also contains free nerve endings that detect the movement of hairs, physical contact, and pain, as well as heat and cold.

TEMPERATURE DETECTORS

To keep its own temperature steady, the body needs to monitor temperature changes around it. It does this with the help of nerve endings in the skin. These respond to cold or heat, and they send signals to the body's temperature control centre in the hypothalamus located in the brain (see p. 122). Skin temperature sensors are best at detecting rapid changes – such as a plunge into icy water taken by the man in the picture on the right. Once the change has taken place, the sensors soon adapt to it.

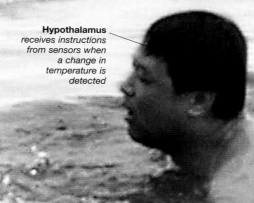

Hypothalamus *receives instructions from sensors when a change in temperature is detected*

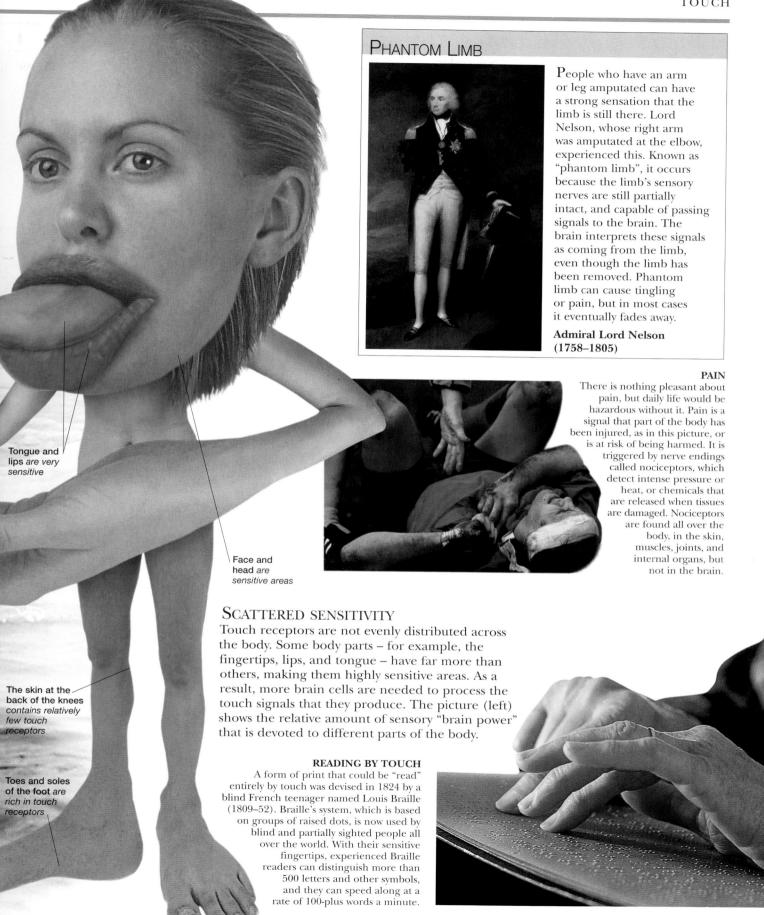

PHANTOM LIMB

People who have an arm or leg amputated can have a strong sensation that the limb is still there. Lord Nelson, whose right arm was amputated at the elbow, experienced this. Known as "phantom limb", it occurs because the limb's sensory nerves are still partially intact, and capable of passing signals to the brain. The brain interprets these signals as coming from the limb, even though the limb has been removed. Phantom limb can cause tingling or pain, but in most cases it eventually fades away.

Admiral Lord Nelson (1758–1805)

PAIN

There is nothing pleasant about pain, but daily life would be hazardous without it. Pain is a signal that part of the body has been injured, as in this picture, or is at risk of being harmed. It is triggered by nerve endings called nociceptors, which detect intense pressure or heat, or chemicals that are released when tissues are damaged. Nociceptors are found all over the body, in the skin, muscles, joints, and internal organs, but not in the brain.

Tongue and lips *are very sensitive*

Face and head *are sensitive areas*

The skin at the back of the knees *contains relatively few touch receptors*

Toes and soles of the foot *are rich in touch receptors*

SCATTERED SENSITIVITY

Touch receptors are not evenly distributed across the body. Some body parts – for example, the fingertips, lips, and tongue – have far more than others, making them highly sensitive areas. As a result, more brain cells are needed to process the touch signals that they produce. The picture (left) shows the relative amount of sensory "brain power" that is devoted to different parts of the body.

READING BY TOUCH

A form of print that could be "read" entirely by touch was devised in 1824 by a blind French teenager named Louis Braille (1809–52). Braille's system, which is based on groups of raised dots, is now used by blind and partially sighted people all over the world. With their sensitive fingertips, experienced Braille readers can distinguish more than 500 letters and other symbols, and they can speed along at a rate of 100-plus words a minute.

Smell and taste

COMPARED WITH THE OTHER special senses, taste and smell are close partners, and work in similar ways. Both are chemical senses: taste detects substances that are dissolved in saliva, while smell detects those present in the air. To do this, they use chemoreceptors, which are specialized cells that respond to specific molecules. Together, these two senses allow us to identify things that are good to eat or drink, but they also warn us about things that might be dangerous. The ability to taste and smell varies a great deal from person to person. This explains why some people make good wine tasters or perfumiers, while others have trouble perceiving some flavours or odours at all.

CHEMICAL DETECTORS

Taste receptors are found on the tongue, while smell receptors are located in the roof of the nasal cavity. Both are triggered by contact with chemicals, but their sensitivity is quite different. Taste receptors can detect only four overall tastes – sweet, salty, sour, and bitter – but smell receptors can distinguish between more than 10,000 different odours. Smell receptors are also much better at detecting faint chemical traces. They can pick up some particularly smelly substances in concentrations of just a few parts per billion, which is why a skunk's scent can be smelt some distance away.

Olfactory bulb *carries nerve impulses to brain from smell receptors*

Olfactory epithelium *contains smell sensors*

Odour molecules *breathed in with air through nostrils*

Taste sensors *located on surface of tongue*

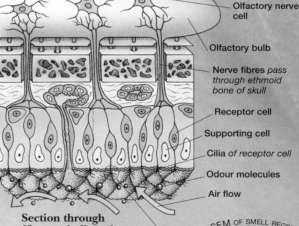

Olfactory nerve cell

Olfactory bulb

Nerve fibres *pass through ethmoid bone of skull*

Receptor cell

Supporting cell

Cilia *of receptor cell*

Odour molecules

Air flow

Section through olfactory bulb and olfactory epithelium

SEM OF SMELL RECEPTORS

SMELL RECEPTORS

Located deep inside the nasal cavity are smell receptors, or olfactory cells, in a thumbnail-sized patch of tissue called the olfactory epithelium. One end of each receptor connects with the olfactory bulb, which is an extension of the brain. The other ends in a cluster of cilia, which look like microscopic hairs. These hairs project into the mucous film lining the nasal cavity, where they respond to dissolved molecules from the incoming air.

Impulses *sent from the eye to the brain give visual clues to aid balance*

The brain *responds to the information it receives by sending out motor impulses to muscles to control movement and thus help balance*

Information *from the balance organs in the ear is sent to the brain*

Impulses *from stretch receptors in the neck muscles indicate which way the head is turned*

Stretch receptors *in the muscles and joints respond as the limbs move, and send impulses to the brain*

Touch receptors *in the feet send impulses to the brain about what sort of surface is underfoot*

STAYING BALANCED

Without balance we would not be able to stand upright. This is because an upright body is naturally unstable – without constant adjustments in posture, it would soon fall down. These adjustments are made by the central nervous system, which receives a constant stream of signals from the organs of the inner ear, as well as from stretch receptors in muscles, and from the eyes. These signals are used to regulate the tension (tone) of skeletal muscles. If the body starts to tip even slightly, groups of muscles tighten to bring it back into line.

MOTION SICKNESS

Sudden movements often make people dizzy, but motion sickness is a longer lasting and much more unpleasant experience. It usually happens when the brain receives conflicting information from the organs of balance and from the eyes. For example, inside a ship on a stormy sea, there are few visual clues that the ship is moving up and down. However, the organs of balance detect these movements, and send signals to the brain. The result is sea sickness – a condition that some people still suffer from, even after years afloat. Cars and planes can also cause motion sickness, but even watching movement at the movies may be enough to trigger it off.

The eye

VISION IS THE FOREMOST of the special senses, and the one that dominates our impressions of the outside world. We use it every moment we are awake, and we dream with visual images when we sleep. Visual expressions crop up almost every time we speak, and when we think of ourselves or other people, it is almost always in a visual way. The organ responsible for triggering this imagery – the eye – is one of the most complex in the body. Set in a protective socket inside the skull, it is always on the move, and is quick to adapt to changing light conditions. It automatically focuses on anything that comes into view, collecting light and converting it into a stream of billions of nervous impulses. Once they have arrived in the brain, these signals have to be analyzed – an awesomely complex process that makes sense of what we see.

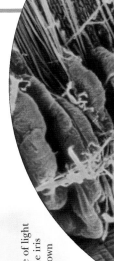

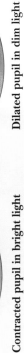

Dilated pupil in dim light

Contracted pupil in bright light

PUPIL SIZE
Human eyes have to cope with a huge range of light intensity. In bright conditions, muscles in the iris contract, narrowing the pupil and cutting down the amount of light entering the eye. If it is dim, the reverse happens. This reflex action takes about a fifth of a second. Pupils also narrow when the eyes look at something very close, because this increases the eye's depth of field.

Eyebrows direct sweat away from the eye and help to keep out some light

The tear gland is under the outer edge of the eyebrow and produces tears

Tears move down and across the eye, carrying debris with them

Pupil is the hole that lets light into the eye

Eyelids protect the eye from bright light and foreign objects; when we blink, they wipe the eyes and keep them clear of dust

Eyelashes stop too much light entering the eye, as well as protecting it from foreign particles

Tear ducts empty tears onto the surface of the eye

Iris colour is determined by the amount of melanin present – brown eyes have the most melanin and blue eyes the least

Tears drain away through two tear openings in the corner of the eye

Nasolacrimal duct drains tears into the nasal cavity

Opening into nasal cavity

Nostril

EYE PROTECTION
Only the iris and pupil of each eyeball – about one sixth – can be seen from the outside. The rest lies inside the eye socket, or orbit, where it is cushioned by pads of fat. Most features surrounding the eye have a protective function – the eyebrows, eyelashes, and eyelids all help shield the eye from excessive light and dust. The front of the eye is kept moist by tears, which wash away specks of dust and help to prevent infection. Tears are washed across the eye by blinking – a reflex action that also helps to protect the eye if anything heads its way.

BANTING AND BEST
Frederick Banting (1899–1978), far right, and Charles Best (1891–1941) isolated insulin. They injected it into Marjorie, a dying diabetic dog whose pancreas had been removed. She recovered. Banting won a Nobel Prize for his work but Best's achievement was overlooked, although Banting shared the prize money with him.

1922. The crucial test with a human patient came in 1923, when they treated Leonard Thompson, a 14-year-old diabetic who was close to death. The effect was miraculous. His blood sugar level fell, one day later he was on his feet, and he soon returned home to lead a normal life with the aid of insulin injections. The Eli Lilly pharmaceutical company immediately began large-scale production of insulin, extracted from pig pancreas. This method has now been superseded by the use of laboratory-produced human insulin. Diabetes is still an incurable disease, but diabetics, once condemned to a short and painful existence, can now live a full and active life.

INSULIN PEN
Some diabetics control their blood glucose by injecting insulin into their blood. The insulin pen gives a measured dose, according to the patient's varying blood glucose levels.

Twenty years earlier, a German scientist, Paul Langerhans (1847–88), had described unique groups of cells within the pancreas, later named islets of Langerhans. British scientist Edward Sharpey-Schafer (1850–1935) showed that a substance regulating glucose metabolism in the body was produced by these islets and called it "insuline" (later shortened to insulin), from the Latin *insula*, meaning island. English physiologist Ernest Starling (1866–1927) coined the name "hormone" in 1905 to describe substances secreted by endocrine glands, like the pancreas, that control metabolism. So now the race was on to isolate the pancreatic hormone, insulin.

Insulin *is a small protein made up of two amino acid chains*

Six insulin molecules *shown in green and purple are bound together here*

COMPUTER-GENERATED IMAGE OF INSULIN MOLECULE

ROLE OF INSULIN
Insulin molecules bind to the outer membrane of cells, triggering chemical processes inside that break down glucose molecules, so removing the sugar from the blood and generating energy. Insulin also inhibits breakdown of glycogen to glucose by the liver.

Insulin production sites *are stained orange*

MAKING INSULIN
Initially, insulin was laboriously extracted from pig pancreas, obtained from slaughterhouses. In some patients, the immune system reacts against pig insulin. This can be avoided by using human insulin, made in the laboratory using *E. coli* bacteria that have human insulin genes inserted into them.

ISOLATION OF INSULIN

Attempts to treat diabetics by feeding them pancreas failed, so extracting the hormone and injecting it into their blood was the only option. After many unsuccessful attempts, Frederick Banting and Charles Best succeeded in isolating insulin in

TEM OF GENETICALLY ALTERED E. COLI BACTERIA

Escherichia coli bacteria *are genetically engineered to produce insulin*

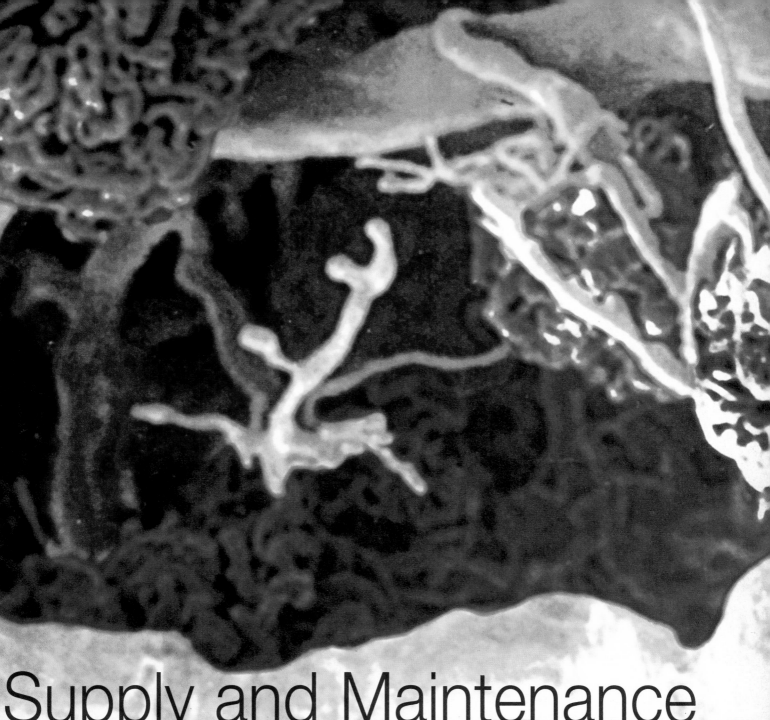

Supply and Maintenance

TO WORK AT THEIR BEST, the body's trillions of cells must be kept in a stable environment, regardless of conditions outside the body. This stability is provided by the systems that supply and maintain body tissues. Together, they ensure that all cells have sufficient food and oxygen to meet their energy requirements, have their wastes removed, are protected from disease-causing pathogens, are kept at a constant, warm temperature of 37°C (98.6°F), and are bathed by a nurturing fluid, the composition of which fluctuates little.

CARDIOVASCULAR system

THROUGHOUT EVERY SECOND of every day of life, as the body's cells perform their incredible variety of functions, they are dependent on a continuous delivery of oxygen and nutrients, and the removal of wastes. These are carried by blood, a tissue made up of cells suspended in fluid that is transported by the cardiovascular, or circulatory, system. Consisting of a muscular pump (the heart) and an intricate network of tubes called blood vessels, the cardiovascular system ensures that blood reaches every part of the body. The amount of blood supplied to the various tissues can be adjusted whenever necessary to meet the body's continually changing needs. In addition to being responsible for transport, the cardiovascular system plays a key role in defending the body against infection through the white blood cells that it carries. It also helps maintain a stable environment within the body.

TRANSPORT NETWORK

The intricate network of blood vessels extends to just about every part of the body. These vessels range in size from the thickness of a thumb to a fraction of a hair's width, visible only under the microscope. So extensive is the network, in fact, that if a person's blood vessels were laid end to end they would stretch out over 100,000 km (62,500 miles). In this "body map" (right), only the major blood vessels are shown, along with the heart, which pumps blood along them. Arteries – vessels carrying blood away from the heart – generally transport oxygen-rich blood (red) to the tissues, while veins – vessels carrying blood to the heart – usually carry oxygen-poor blood (blue). Connecting arteries and veins – but not visible here – is a massive system of microscopically narrow capillaries.

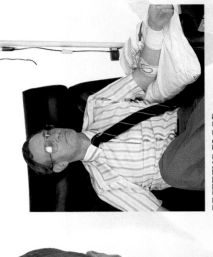

RECEIVING BLOOD
A blood transfusion can be given to restore blood volume. This can be life-saving if blood loss is severe. Blood is grouped according to the presence of certain proteins on the surface of red blood cells. The blood to be transfused and the recipient's blood must be "typed" to ensure they are compatible.

Pulmonary arteries carry oxygen-poor blood from the right side of the heart to the lungs

Oxygen-poor blood (blue)

Pulmonary veins carry oxygen-rich blood from the lungs to the left side of the heart

The heart (shown here cut open) is a muscular pump that propels blood around the circulatory system

Inferior vena cava is the main vein carrying blood from the abdomen and legs to the heart

Aorta is the main artery through which blood leaves the heart

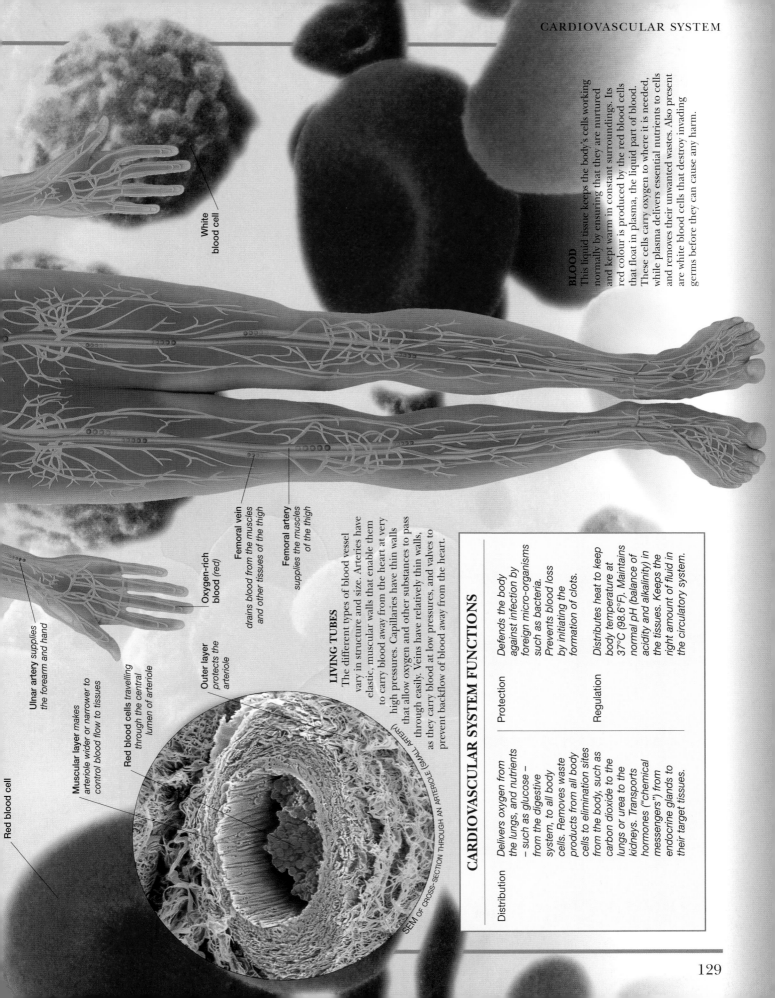

White blood cell

Femoral vein
*drains blood from the muscles
and other tissues of the thigh*

Femoral artery
*supplies the muscles
of the thigh*

Oxygen-rich
blood (red)

Outer layer
*protects the
arteriole*

Red blood cells travelling
*through the central
lumen of arteriole*

Muscular layer makes
*arteriole wider or narrower to
control blood flow to tissues*

Ulnar artery supplies
the forearm and hand

Red blood cell

SEM OF CROSS-SECTION THROUGH AN ARTERIOLE (SMALL ARTERY)

BLOOD

This liquid tissue keeps the body's cells working normally by ensuring that they are nurtured and kept warm in constant surroundings. Its red colour is produced by the red blood cells that float in plasma, the liquid part of blood. These cells carry oxygen to where it is needed, while plasma delivers essential nutrients to cells and removes their unwanted wastes. Also present are white blood cells that destroy invading germs before they can cause any harm.

LIVING TUBES

The different types of blood vessel vary in structure and size. Arteries have elastic, muscular walls that enable them to carry blood away from the heart at very high pressures. Capillaries have thin walls that allow oxygen and other substances to pass through easily. Veins have relatively thin walls, as they carry blood at low pressures, and valves to prevent backflow of blood away from the heart.

CARDIOVASCULAR SYSTEM FUNCTIONS

Distribution	*Delivers oxygen from the lungs, and nutrients – such as glucose – from the digestive system, to all body cells. Removes waste products from all body cells to elimination sites from the body, such as carbon dioxide to the lungs or urea to the kidneys. Transports hormones ("chemical messengers") from endocrine glands to their target tissues.*
Protection	*Defends the body against infection by foreign micro-organisms such as bacteria. Prevents blood loss by initiating the formation of clots.*
Regulation	*Distributes heat to keep body temperature at 37°C (98.6°F). Maintains normal pH (balance of acidity and alkalinity) in the tissues. Keeps the right amount of fluid in the circulatory system.*

Blood

COURSING CONTINUOUSLY through the network of blood vessels, blood supplies the needs of each of the body's trillions of cells, ensuring that they are kept in safe, constant surroundings. It fulfils this demanding role in three ways. Firstly, it acts as the body's delivery and collection service, carrying food, oxygen, hormones, and other essentials to cells, and removing wastes. Secondly, it distributes heat around the body, keeping cells at a steady 37°C (98.6°F). Thirdly, it helps defend the body against infection. Blood itself is not a simple liquid but consists of different types of cells floating in fluid plasma. Each component – whether cellular or liquid – has a particular role within the body.

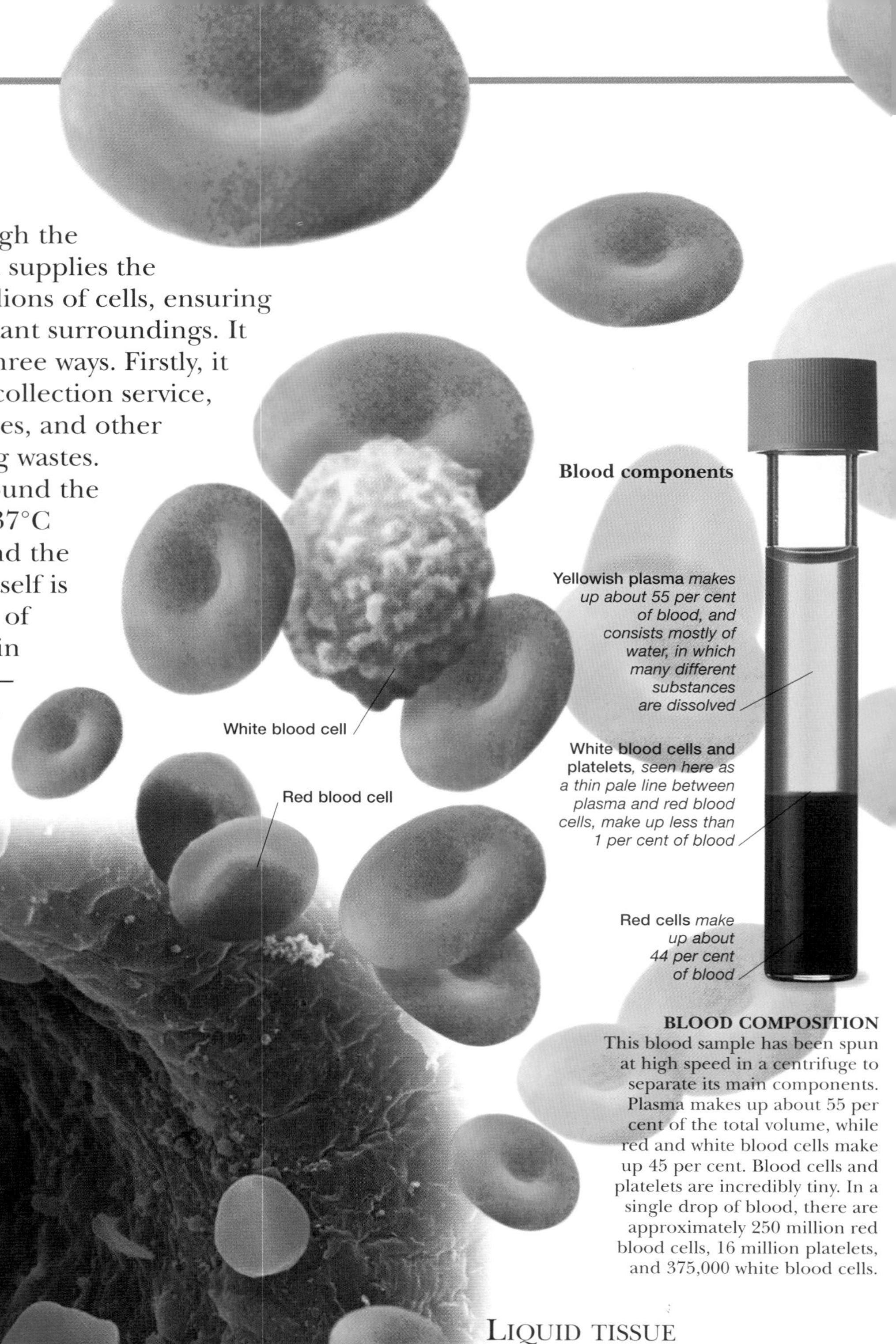

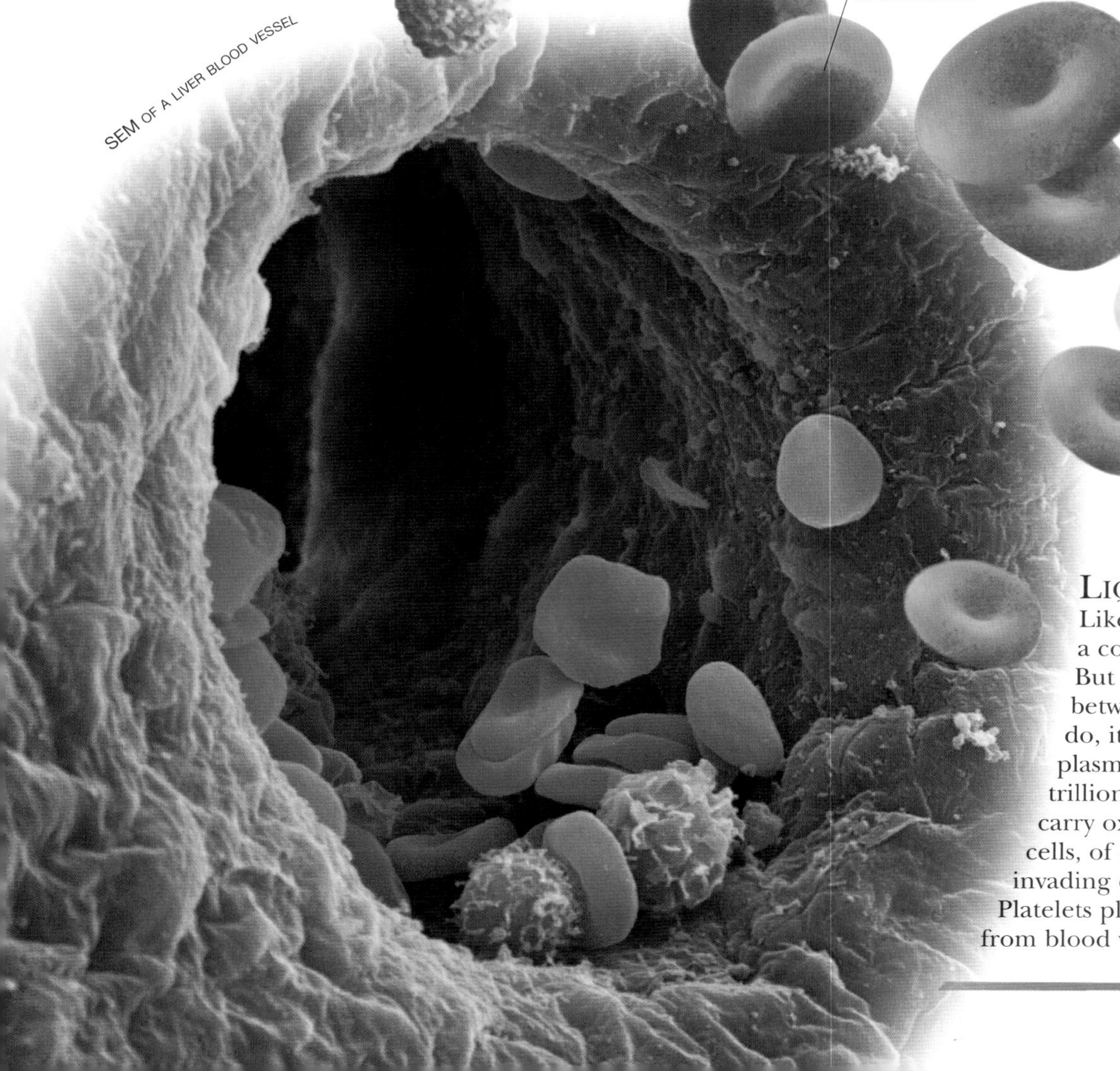

SEM OF A LIVER BLOOD VESSEL

White blood cell

Red blood cell

Blood components

Yellowish plasma *makes up about 55 per cent of blood, and consists mostly of water, in which many different substances are dissolved*

White blood cells and platelets, *seen here as a thin pale line between plasma and red blood cells, make up less than 1 per cent of blood*

Red cells *make up about 44 per cent of blood*

BLOOD COMPOSITION
This blood sample has been spun at high speed in a centrifuge to separate its main components. Plasma makes up about 55 per cent of the total volume, while red and white blood cells make up 45 per cent. Blood cells and platelets are incredibly tiny. In a single drop of blood, there are approximately 250 million red blood cells, 16 million platelets, and 375,000 white blood cells.

LIQUID TISSUE

Like bone and cartilage, blood is a connective tissue (see pp. 26–7). But instead of having a solid matrix between its cells as the other tissues do, it has a liquid matrix called plasma. Suspended in plasma are trillions of blood cells. Red blood cells carry oxygen to the tissues. White blood cells, of different types, detect and destroy invading disease-causing micro-organisms. Platelets play a key role in stopping leaks from blood vessels by making blood clot.

PLASMA

This straw-coloured liquid consists of about 90 per cent water, with the remaining 10 per cent made up of some 100 solutes (dissolved substances) that are either carried by the blood or help maintain its stable composition. Key solutes are listed (far right) with a description of their function or delivery destination.

Transfusion bag containing plasma

SUBSTANCES DISSOLVED IN PLASMA	FUNCTION
Proteins	Plasma proteins have various functions:
• albumin	Maintains water balance between blood and tissues, and helps regulate blood volume
• alpha and beta globulins	Helps transport some lipids, vitamins, and hormones
• gamma globulins	Antibodies that help defend the body against pathogens
• clotting proteins, including fibrinogen	Play an essential role in blood clotting
• others	Metabolic enzymes and antibacterial proteins
Nutrients: e.g. glucose, amino acids, and fatty acids	Products of digestion carried from the small intestine to be used by cells, stored, or broken down
Carbon dioxide	Waste product of cell respiration carried to lungs for disposal
Nitrogenous wastes	Products of protein breakdown, including urea carried from liver to kidneys and other points of excretion
Hormones	Chemical messengers carried from endocrine glands to target tissues
Electrolytes (ions)	Including sodium and potassium ions, which help maintain normal blood concentration

HOW MUCH?

Blood makes up about 8 per cent of body weight. This amounts to 4–5 litres (7–8.8 pints) in women and 5–6 litres (8.8–10.6 pints) in men. Blood volume is regulated by the kidneys (see pp. 206–7), which increase or decrease the amount of fluid they filter from blood and pass out of the body as urine.

Blood cells *are made in the parts of the skeleton shaded red*

The average person has 5 litres (8.8 pints) of blood

Ten of these 500 ml (0.9 pint) bottles *contain the same amount of blood as an average person*

White blood cell

Red blood cells

Platelet

SEM OF BONE MARROW

Sinusoid *channel that carries blood cells*

MAKING BLOOD CELLS

Red blood cells, platelets, and some white blood cells are made by the red bone marrow found in the flat bones of the axial skeleton – cranium, collar bones, shoulder blades, breast bone, ribs, vertebrae, hip bones – and the top of each humerus (upper arm bone) and femur (thigh bone). All cell types arise from the same basic stem cells, called haemocytoblasts ("blood cell sprouts"), but each type of cell follows a different development pathway.

Red blood cells

THE AVERAGE ADULT HAS around 25 trillion red blood cells in their blood at any one time. Also known as erythrocytes, these tiny cells make up 99 per cent of the cells in the blood and perform the vital function of carrying oxygen to all the tissues of the body. They are produced at a rate of over 2 million per second in the red marrow of certain bones. Their unique structure means they are able to squeeze through the tiniest of blood vessels and can pick up oxygen readily in the lungs and unload it in the tissues. In addition to transporting oxygen, they also carry some of the carbon dioxide produced as a waste product by the cells of the body.

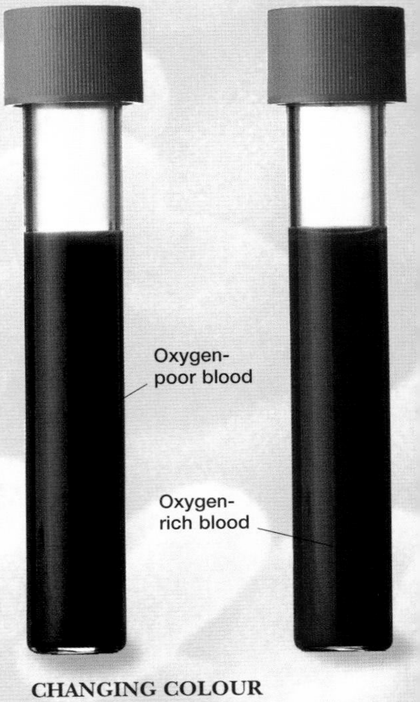

Oxygen-poor blood

Oxygen-rich blood

LARGE SURFACES
As this cutaway view shows, a red blood cell resembles a flattened, dimpled disc, a shape that gives the cell a large surface area in relation to its volume. It means that no oxygen-carrying haemoglobin molecule is ever far from the cell membrane, which explains why red blood cells are very efficient at both picking up and depositing oxygen.

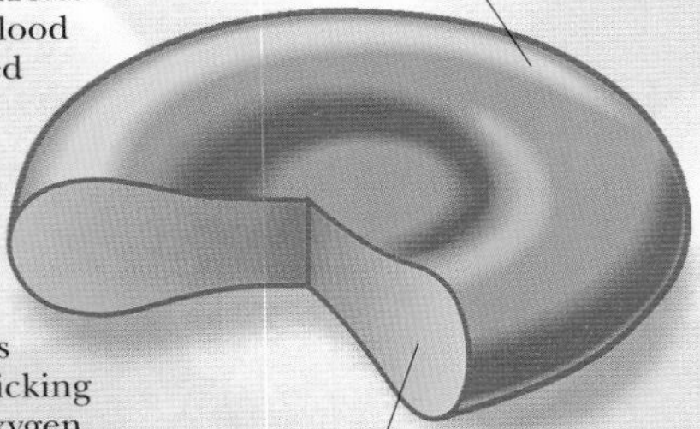

Cell membrane

Haemoglobin *packs the interior of the cell*

Cut-away of red blood cell

CHANGING COLOUR
The redness of blood alters as it travels around the body because haemoglobin changes its colour as it picks up or deposits oxygen. After haemoglobin picks up oxygen in the lungs, oxygen-rich blood travelling along the arteries to the tissues is bright red in colour. After haemoglobin unloads oxygen in the tissues, the oxygen-poor blood returning along veins to the heart is now a dull, dark red. This gives thin-walled veins just under the skin's surface a blue appearance.

OXYGEN CARRIERS
Red blood cells have a number of special features, including their dimpled shape and flexibility. They are also the only cells in the body that do not have a nucleus. Instead, they are packed with haemoglobin, a protein that is responsible for their red colour, and for the redness of the blood itself. Haemoglobin molecules pick up oxygen in the lungs and release it in the tissues. The formation and destruction of red blood cells takes place continuously. They begin their life as immature stem cells in the bone marrow, becoming fully developed and capable of transporting oxygen over a period of about five days. Each red blood cell lasts for about 120 days.

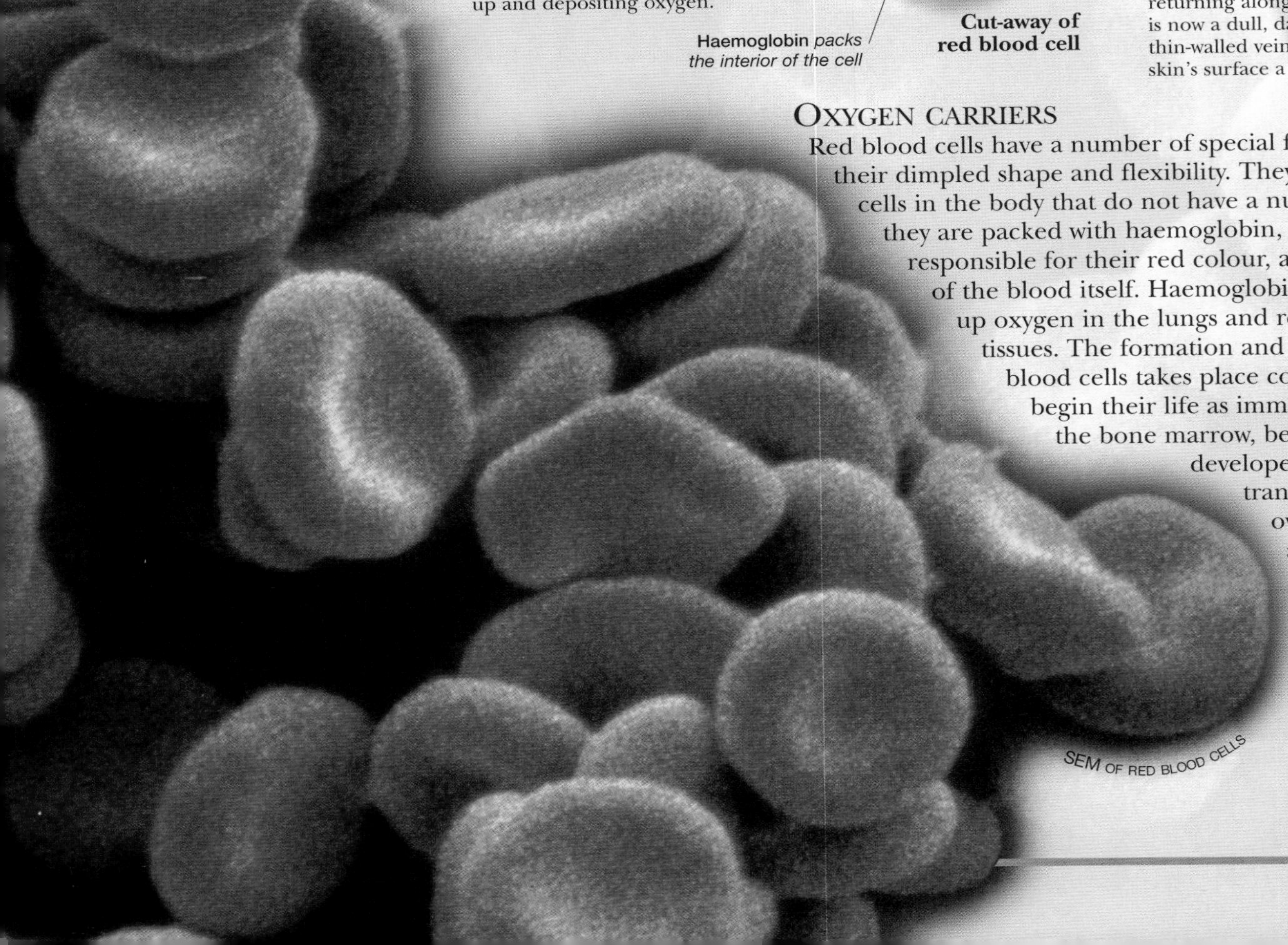

SEM OF RED BLOOD CELLS

BLOOD GROUP COMPATIBILITY

Blood is always typed before transfusion to make sure a person receives blood that matches their own ABO and Rh blood groups to avoid adverse reactions. In an emergency, however, group O blood can be given to anyone, while someone who is group AB can receive blood from any donor. This chart shows compatible donors and recipients.

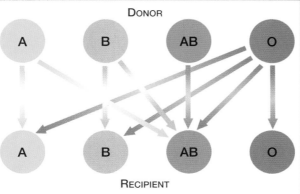

DONOR

A   B   AB   O

A   B   AB   O

RECIPIENT

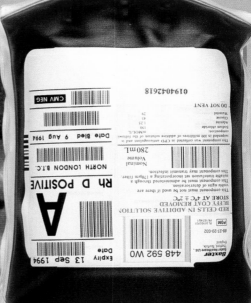

BLOOD BANK
Blood is tested for diseases and typed for its blood group. Chemicals are added to increase its shelf life, and it is stored at low temperature in hospitals until it is needed. It can be stored either as blood, or as one of its components, such as plasma.

blood vessels. The explanation for this was provided by Karl Landsteiner (1868–1943), who in 1901 discovered that within the human population there were different blood groups. Which of the four blood groups a person belonged to – A, B, AB, or O – depended on the presence or absence of two molecular markers called antigens – identified as A and B – carried on the surface of their red blood cells. Group A has the A antigen, group B the B antigen, group AB both antigens, and group O neither antigen. Also present in blood plasma are antibodies that act against antigens which are not present on a person's own red blood cells. Group A blood contains anti-B antibodies, group B blood contains anti-A antibodies, group O blood contains both antibodies, and group AB blood contains neither. This explains why, for example, if a group A person is given group B blood, their anti-B antibodies "recognize" the B antigens on the foreign red blood cells, make them stick together, and block blood vessels.

In 1940, Landsteiner recognized the rhesus antigen (Rh), first identified in rhesus monkeys. About 85 per cent of people are rhesus positive (Rh+) because they carry the antigen, while the remainder lack the antigen and are therefore rhesus negative (Rh–).

SAFE TRANSFUSIONS

Once an individual's blood group could be identified by simple tests, safe blood transfusions became routine. But obtaining enough blood for emergency use is still a problem. Blood can be stored at low temperature in blood banks, but only for limited periods, so an important long-term goal is to develop artificial blood that can be given to a member of any blood group.

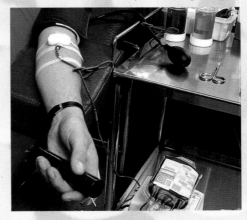

GIVING BLOOD
Healthy individuals can donate 500 ml (0.9 pint) of blood every 16 weeks. They can also donate parts of their blood, such as red cells, white cells, plasma, or platelets, more frequently. To give blood, the donor lies down and has a needle inserted into a vein near the elbow. Blood flows down a plastic tube and into a storage bag.

Heart

ONCE THOUGHT TO BE the source of feelings of love and emotion, the heart is actually the powerhouse of the circulatory system. Rhythmic contractions of this muscular pump push blood along the blood vessels to all parts of the body, even its far extremities, and back to the heart again. The beating heart ensures that every cell of the body has an uninterrupted supply of food, oxygen, and other essentials. So powerful is the heart that it can pump the body's entire blood volume of 5 litres (8.8 pints) around the body about once every minute. On average it beats, or pumps, 70 times a minute when the body is at rest, yet can increase this rate if the body is more active. Over a lifetime of 70 years, the heart beats some 2.5 billion times without tiring or stopping for a rest, thanks to the cardiac muscle in its walls.

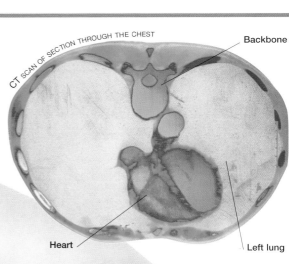

CT SCAN OF SECTION THROUGH THE CHEST

Backbone

Heart

Left lung

HEART AND LUNGS
This CT scan "slices" through the chest to show clearly how the heart is surrounded by right and left lungs. Being this close is important. It cuts to a minimum the distance blood has to travel from the heart to pick up oxygen in the lungs.

TWO SIDES
Divided into the left and right sides, the heart is two pumps in one. On each side, blood enters the atrium, then passes into the ventricle to be pumped on its onward journey. On the right side, oxygen-poor blood (blue) enters the right atrium, flows into the right ventricle, and is pumped to the lungs. On the left side, oxygen-rich blood (red) enters the left atrium, flows into the left ventricle, and is pumped to the rest of the body.

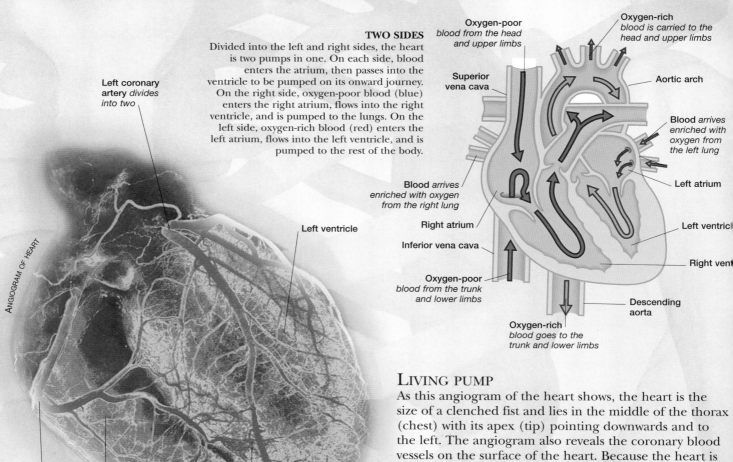

ANGIOGRAM OF HEART

Left coronary artery *divides into two*

Left ventricle

Right ventricle

Right coronary artery

Network of blood vessels *supplies heart muscle with oxygen and food*

Heart *is tilted towards the left side of the body*

Oxygen-poor *blood from the head and upper limbs*

Superior vena cava

Oxygen-rich *blood is carried to the head and upper limbs*

Aortic arch

Blood *arrives enriched with oxygen from the left lung*

Left atrium

Left ventricle

Right ventricle

Descending aorta

Blood *arrives enriched with oxygen from the right lung*

Right atrium

Inferior vena cava

Oxygen-poor *blood from the trunk and lower limbs*

Oxygen-rich *blood goes to the trunk and lower limbs*

LIVING PUMP

As this angiogram of the heart shows, the heart is the size of a clenched fist and lies in the middle of the thorax (chest) with its apex (tip) pointing downwards and to the left. The angiogram also reveals the coronary blood vessels on the surface of the heart. Because the heart is a pump made out of living tissue, it too requires a supply of blood. However, the blood that gushes through it every second cannot meet the demands of the heart muscle for food and oxygen. To overcome this problem, coronary arteries arise from the aorta as it leaves the heart, and branch to carry oxygen-rich blood to the cardiac muscle in the heart wall (see also pp. 144–5).

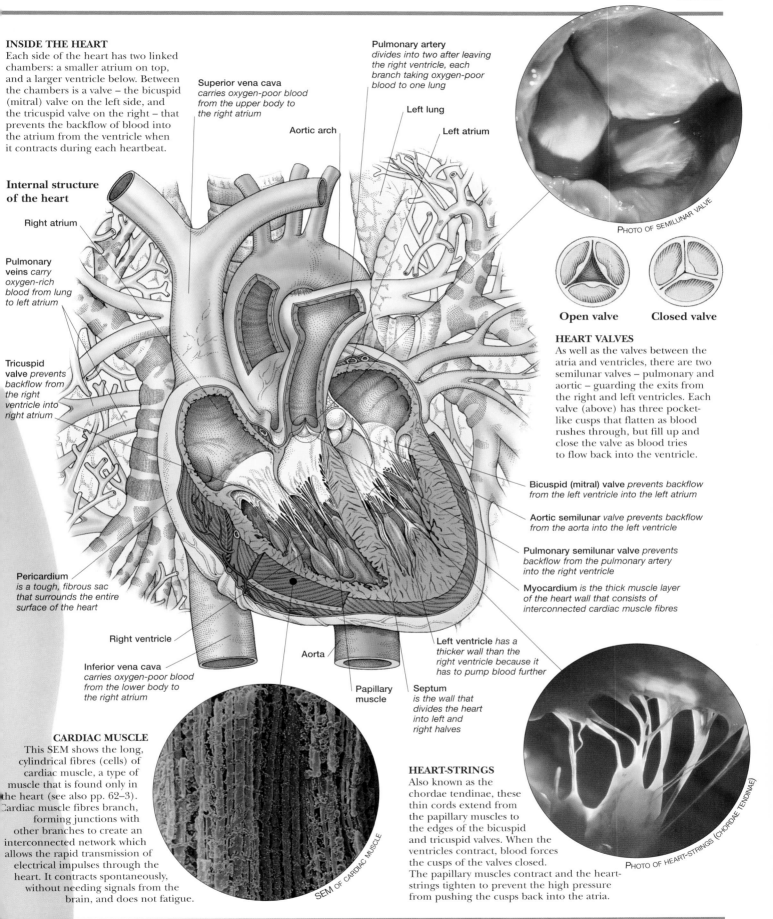

INSIDE THE HEART
Each side of the heart has two linked chambers: a smaller atrium on top, and a larger ventricle below. Between the chambers is a valve – the bicuspid (mitral) valve on the left side, and the tricuspid valve on the right – that prevents the backflow of blood into the atrium from the ventricle when it contracts during each heartbeat.

Internal structure of the heart

Right atrium

Pulmonary veins *carry oxygen-rich blood from lung to left atrium*

Tricuspid valve *prevents backflow from the right ventricle into right atrium*

Pericardium *is a tough, fibrous sac that surrounds the entire surface of the heart*

Right ventricle

Inferior vena cava *carries oxygen-poor blood from the lower body to the right atrium*

Superior vena cava *carries oxygen-poor blood from the upper body to the right atrium*

Aortic arch

Pulmonary artery *divides into two after leaving the right ventricle, each branch taking oxygen-poor blood to one lung*

Left lung

Left atrium

Aorta

Papillary muscle

Septum *is the wall that divides the heart into left and right halves*

Left ventricle *has a thicker wall than the right ventricle because it has to pump blood further*

PHOTO OF SEMILUNAR VALVE

Open valve   **Closed valve**

HEART VALVES
As well as the valves between the atria and ventricles, there are two semilunar valves – pulmonary and aortic – guarding the exits from the right and left ventricles. Each valve (above) has three pocket-like cusps that flatten as blood rushes through, but fill up and close the valve as blood tries to flow back into the ventricle.

Bicuspid (mitral) valve *prevents backflow from the left ventricle into the left atrium*

Aortic semilunar *valve prevents backflow from the aorta into the left ventricle*

Pulmonary semilunar valve *prevents backflow from the pulmonary artery into the right ventricle*

Myocardium *is the thick muscle layer of the heart wall that consists of interconnected cardiac muscle fibres*

CARDIAC MUSCLE
This SEM shows the long, cylindrical fibres (cells) of cardiac muscle, a type of muscle that is found only in the heart (see also pp. 62–3). Cardiac muscle fibres branch, forming junctions with other branches to create an interconnected network which allows the rapid transmission of electrical impulses through the heart. It contracts spontaneously, without needing signals from the brain, and does not fatigue.

SEM OF CARDIAC MUSCLE

HEART-STRINGS
Also known as the chordae tendinae, these thin cords extend from the papillary muscles to the edges of the bicuspid and tricuspid valves. When the ventricles contract, blood forces the cusps of the valves closed. The papillary muscles contract and the heart-strings tighten to prevent the high pressure from pushing the cusps back into the atria.

PHOTO OF HEART-STRINGS (CHORDAE TENDINAE)

Heartbeats

MOST PEOPLE HAVE experienced the feeling of a pounding heart, especially after running fast and then stopping suddenly. Every rhythmic pulsation represents one heartbeat, but not just one event. Each beat is made up of three stages called the heartbeat cycle. The entire cycle is masterminded by a patch of modified cardiac muscle called the sinoatrial (SA) node that acts as a pacemaker, sending out regular electrical impulses to stimulate contraction of the heart's chambers. While heart contraction happens spontaneously, outside control is needed, in the form of the autonomic nervous system (ANS), which alters the rate and strength of heartbeats in order to meet the ever-changing needs of the body.

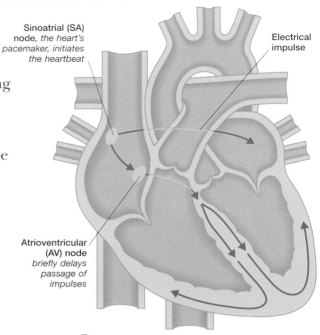

Sinoatrial (SA) node, the heart's pacemaker, initiates the heartbeat

Electrical impulse

Atrioventricular (AV) node briefly delays passage of impulses

HEARTBEAT CYCLE

The three diagrams below show the stages of a heartbeat cycle. During diastole (relaxation), the heart relaxes, the atria fill with blood, and the pulmonary and aortic semilunar valves close to prevent backflow. During atrial systole (contraction), the atria contract to push blood into their respective ventricles. During ventricular systole, as the two ventricles contract together to pump blood out of the heart, blood pressure forces the semilunar valves open and the bicuspid and tricuspid valves shut. The three stages correspond to different sections of an ECG recording, an example of which extends across these pages and is explained (right).

ELECTRICAL PATHWAYS

Electrical impulses spread through the atria from the SA node, then are briefly delayed at the atrioventricular (AV) node, giving the atria time to contract before impulses spread through the ventricles and trigger their contraction. This electrical activity, recorded as an electrocardiogram (ECG), below, shows characteristic peaks. The P wave occurs as impulses travel over the atria, the QRS peak as impulses pass through the ventricles, and the T wave as the ventricles relax.

ECG recording   Stage 1   Stage 2   Stage 3

R

P

Q

S

T

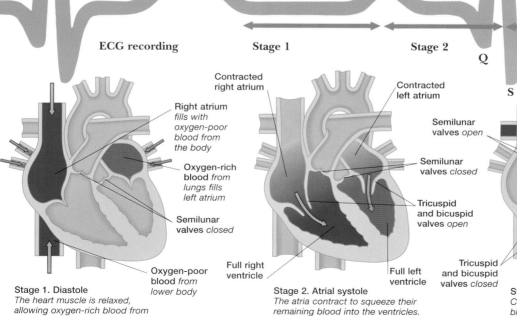

Contracted right atrium

Right atrium fills with oxygen-poor blood from the body

Oxygen-rich blood from lungs fills left atrium

Semilunar valves closed

Oxygen-poor blood from lower body

Contracted left atrium

Semilunar valves open

Semilunar valves closed

Tricuspid and bicuspid valves open

Full right ventricle

Full left ventricle

Tricuspid and bicuspid valves closed

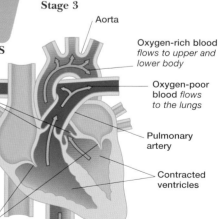

Aorta

Oxygen-rich blood flows to upper and lower body

Oxygen-poor blood flows to the lungs

Pulmonary artery

Contracted ventricles

Stage 1. Diastole
The heart muscle is relaxed, allowing oxygen-rich blood from the lungs and oxygen-poor blood from the body to enter the left and right atria respectively.

Stage 2. Atrial systole
The atria contract to squeeze their remaining blood into the ventricles. The tricuspid and bicuspid valves open to allow this, but the semilunar valves remain closed.

Stage 3. Ventricular systole
Contraction of the ventricles sends blood to the lungs and around the body. The semilunar valves are pushed open by the surge of blood, while the tricuspid and bicuspid valves are closed.

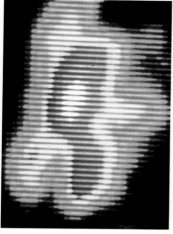

Left ventricle contracted   **Left ventricle relaxed**

HEART SOUNDS

The gamma camera scans (above) provide one way for doctors to follow the heartbeat cycle in action. Another is to use a stethoscope to listen for the sounds produced when the heart's valves close. Each heartbeat produces two sounds: a longer, louder "lub" sound when bicuspid and tricuspid valves close, and a shorter, sharper "dub" sound when the semilunar valves close. Unusual heart sounds can indicate that a valve is leaking.

LISTENING TO HEARTBEATS

The effective working of the valves can be assessed by listening to the sounds they make as they close. Early physicians listened to the heartbeat by placing an ear to the chest, but a one-ear stethoscope, invented by French doctor René Laënnec (1781–1826) in 1816, made the sounds much clearer and easier to hear. About 40 years later, the principles of this instrument were used to develop the two-ear, or binaural, stethoscope, which is still used today. In addition to assessing the heart, a stethoscope can be used to listen to breathing sounds made by the lungs and to sounds generated by the intestines.

Laënnec-type stethoscope c. 1820

CONTROLLING HEART RATE

Heart rate is controlled by the sympathetic and parasympathetic sections of the ANS (see pp. 98—9), whose nerve fibres terminate in the heart's pacemaking SA node. Under orders from the cardioregulatory centre in the brain stem, which monitors conditions inside the body, sympathetic signals speed up heart rate during exercise or stress, while parasympathetic signals slow heart rate when the body is at rest to about 70 beats per minute.

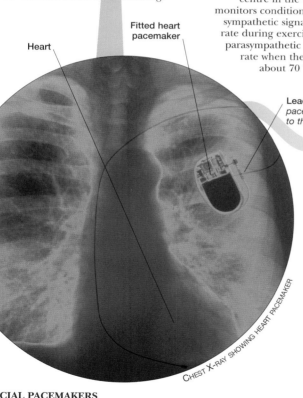

Heart

Fitted heart pacemaker

Lead *connects pacemaker to the heart*

CHEST X-RAY SHOWING HEART PACEMAKER

Sympathetic *speeds up heart rate during exercise*

Parasympathetic *slows heart rate after exercise*

Exercise can increase heart rate significantly

ARTIFICIAL PACEMAKERS

This X-ray shows an artificial pacemaker implanted under the skin of a person whose SA node has stopped working normally, or whose heart does not conduct electrical impulses properly. Powered by a long-life battery, the pacemaker sends electrical impulses along a wire to stimulate the heart to beat. Some send out impulses at a fixed rate, others switch on if the heart misses a beat, while others vary their rate to match body activity levels.

HEART PROBLEMS

U NSEEN INSIDE THE CHEST, the heart is taken for granted until something goes wrong. A common cause of heart problems is a narrowing or blocking of the coronary arteries, which provide heart muscle with oxygen, a condition known as coronary artery disease. Its main symptom is chest pain, noticed during stress or exercise, when extra demands are put on the heart. The chances of developing this problem are increased by smoking, high blood pressure, a high-fat diet, obesity, and inactivity. But before looking at the consequences and treatment of this disease, consider a drastic measure that can deal with a badly damaged heart.

FIRST HEART TRANSPLANTS
In 1967, Christiaan Barnard led a team of 20 surgeons to perform the first human heart transplant.

HEART TRANSPLANTS

First performed in 1967 in South Africa by Professor Christiaan Barnard (1922–2001), a heart transplant involves the replacement of a diseased heart with a healthy one. Before the damaged heart is removed, the recipient's major blood vessels are connected to a bypass machine that pumps oxygen-rich blood to the brain and other organs. Once in place, the healthy heart is connected to the recipient's vessels and the bypass machine is switched off. Rejection of the new heart by the recipient's immune system is a major risk, and the patient must take preventative medication.

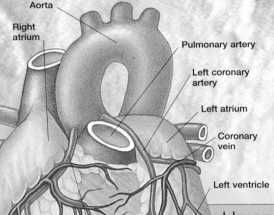

Aorta
Right atrium
Pulmonary artery
Left coronary artery
Left atrium
Coronary vein
Left ventricle
Right ventricle
Right coronary artery

Front view of heart

CORONARY VESSELS
Both main coronary arteries originate from the aorta. The right coronary artery branches to deliver blood mainly to the right side of the heart; the left artery and its branches supply both ventricles and the left atrium, as well as the septum, the muscular wall that divides the heart. The smallest arterial branches connect to the smallest veins by capillaries. Coronary veins return oxygen-poor blood to the coronary sinus, a large vein at the back of the heart that empties into the right atrium.

HEART ATTACKS

Narrowing of coronary arteries can occur as a result of atheroma, a common condition in which fatty plaques are deposited on the lining of arterial walls, causing scarring. If a narrowed artery becomes blocked by a blood clot, the blood supply to an area of the heart is cut off, causing permanent damage to this area. This occurrence is known as a heart attack or myocardial infarction.

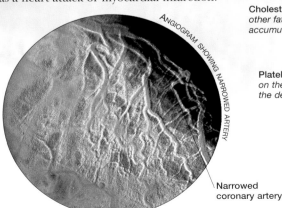

ANGIOGRAM SHOWING NARROWED ARTERY

Narrowed coronary artery

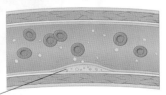

Cholesterol *and other fatty substances accumulate*

EARLY ATHEROSCLEROSIS
The first signs of atherosclerosis, the narrowing of arteries, are the gradual accumulation of fatty substances in the artery wall.

Platelets *collect on the surface of the deposits*

ADVANCED STAGES
The yellow deposits, called atheroma, cause the muscle layer to thicken. They restrict blood flow and narrow the artery.

HEALTHY HEART

The well-being of the heart can be assessed by performing a walking electrocardiogram (ECG), which records the electrical activity within the heart during a period of exercise on a treadmill. The speed of walking is gradually increased – and the treadmill may also be inclined – so that the stress on the heart slowly rises. Some conditions, such as coronary artery disease, may be diagnosed if particular changes are seen on the trace.

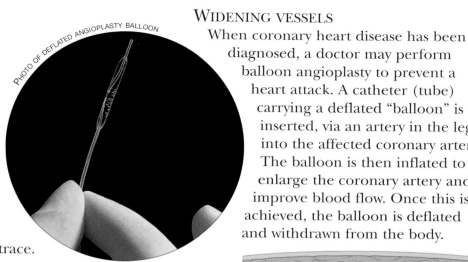
PHOTO OF DEFLATED ANGIOPLASTY BALLOON

WIDENING VESSELS

When coronary heart disease has been diagnosed, a doctor may perform balloon angioplasty to prevent a heart attack. A catheter (tube) carrying a deflated "balloon" is inserted, via an artery in the leg, into the affected coronary artery. The balloon is then inflated to enlarge the coronary artery and improve blood flow. Once this is achieved, the balloon is deflated and withdrawn from the body.

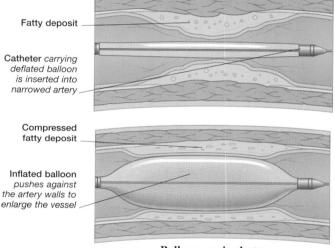

Fatty deposit

Catheter *carrying deflated balloon is inserted into narrowed artery*

Compressed fatty deposit

Inflated balloon *pushes against the artery walls to enlarge the vessel*

Balloon angioplasty

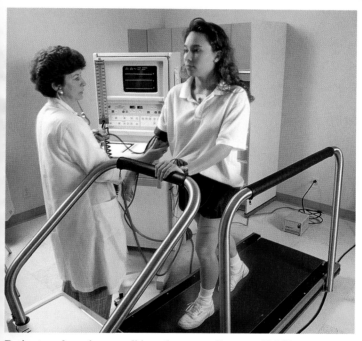

Patient undergoing a walking electrocardiogram (ECG)

EMERGENCY DEFIBRILLATION

Coronary heart disease may result in a heart attack which can cause fibrillation. This occurs when cardiac muscle fibres contract individually and chaotically rather than together, as happens ordinarily. As a consequence, the ventricles cannot pump blood, a condition known as cardiac arrest. To counteract this problem, emergency defibrillation, or cardioversion ("turning the heart"), can be performed by giving the heart a brief electric shock through two metal plates applied to the chest. The shock may restore normal synchronized contractions.

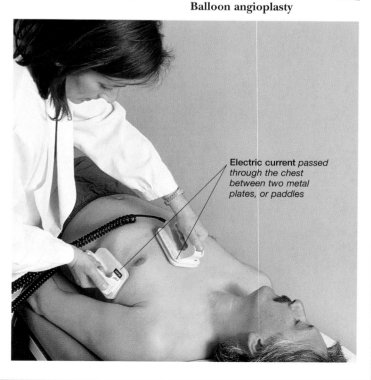

Electric current *passed through the chest between two metal plates, or paddles*

Blood circulation

INSIDE A HUMAN BODY is a system of blood-carrying tubes that, if stretched out, would extend over 150,000 km (93,206 miles), the equivalent of being wrapped around the Earth four times. Some 98 per cent of this incredible distance is made up of the microscopic capillaries that permeate every nook and cranny of the body's tissues to ensure that every cell receives its supplies of food and oxygen and that its waste is removed. Together, the capillaries and other blood vessels – the arteries and veins – form a blood circulation network that follows a figure-of-eight path, first through the lungs to pick up oxygen and then round the body to deliver that oxygen.

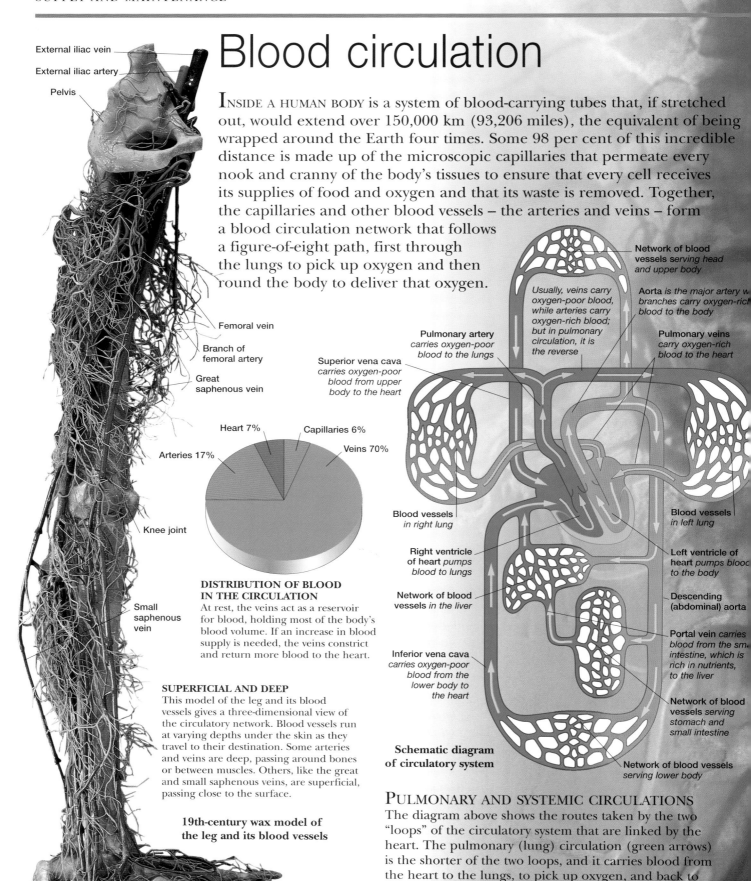

External iliac vein

External iliac artery

Pelvis

Femoral vein

Branch of femoral artery

Great saphenous vein

Knee joint

Small saphenous vein

Bones of the foot

Heart 7%

Capillaries 6%

Arteries 17%

Veins 70%

DISTRIBUTION OF BLOOD IN THE CIRCULATION
At rest, the veins act as a reservoir for blood, holding most of the body's blood volume. If an increase in blood supply is needed, the veins constrict and return more blood to the heart.

SUPERFICIAL AND DEEP
This model of the leg and its blood vessels gives a three-dimensional view of the circulatory network. Blood vessels run at varying depths under the skin as they travel to their destination. Some arteries and veins are deep, passing around bones or between muscles. Others, like the great and small saphenous veins, are superficial, passing close to the surface.

19th-century wax model of the leg and its blood vessels

Network of blood vessels *serving head and upper body*

Aorta *is the major artery w branches carry oxygen-rich blood to the body*

Usually, veins carry oxygen-poor blood, while arteries carry oxygen-rich blood; but in pulmonary circulation, it is the reverse

Pulmonary artery *carries oxygen-poor blood to the lungs*

Pulmonary veins *carry oxygen-rich blood to the heart*

Superior vena cava *carries oxygen-poor blood from upper body to the heart*

Blood vessels *in right lung*

Blood vessels *in left lung*

Right ventricle of heart *pumps blood to lungs*

Left ventricle of heart *pumps blood to the body*

Network of blood vessels *in the liver*

Descending (abdominal) aorta

Portal vein *carries blood from the sm intestine, which is rich in nutrients, to the liver*

Inferior vena cava *carries oxygen-poor blood from the lower body to the heart*

Network of blood vessels *serving stomach and small intestine*

Schematic diagram of circulatory system

Network of blood vessels *serving lower body*

PULMONARY AND SYSTEMIC CIRCULATIONS

The diagram above shows the routes taken by the two "loops" of the circulatory system that are linked by the heart. The pulmonary (lung) circulation (green arrows) is the shorter of the two loops, and it carries blood from the heart to the lungs, to pick up oxygen, and back to the heart. The systemic (body) circulation (yellow arrows) carries blood to all body tissues and back to the heart.

MEASURING BLOOD PRESSURE

Blood pressure is measured using a sphygmomanometer (literally a "pulse pressure measurer"), which consists of an inflatable cuff linked to a pressure gauge, and a stethoscope. In fact, two pressures, not one, are measured (see below) – systolic and then diastolic. The cuff is wrapped around the upper arm and inflated to squeeze the arm until blood flow along the brachial artery (see p. 149) stops. Placing the end of the stethoscope on the skin over the brachial artery, the doctor now deflates the cuff until a pulse can just be heard. The reading on the gauge shows the higher, systolic pressure, which is sufficient to push blood along the narrowed artery. The cuff is deflated further until the pulse sound just disappears, indicating that blood is flowing freely along the artery. The gauge now shows the lower, diastolic pressure.

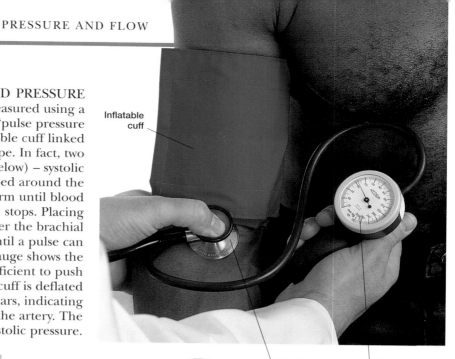

Inflatable cuff

A stethoscope is used by a doctor to listen to blood flow

Gauge indicates pressure

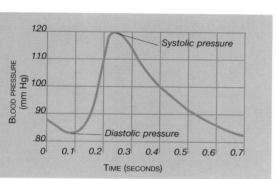

SYSTOLIC AND DIASTOLIC

During each heartbeat cycle (see pp. 142–3), the heart contracts (systole), causing a peak in arterial blood pressure called systolic pressure (see graph, above), then relaxes (diastole), causing a fall to the minimum, diastolic pressure. Pressure, measured in millimetres of mercury (mm Hg), varies according to age, sex, and health, but in a healthy young adult should be about 120/80 (120 mm Hg systolic and 80 mm Hg diastolic).

FROSTBITE

At temperatures below 0˚C (32˚F), the small arteries supplying the skin and underlying tissues narrow, restricting the supply of blood. If the cold conditions persist, ice is deposited in the tissues, causing damage. The fingers, toes, and nose are particularly susceptible to this condition, known as frostbite. The damage may be permanent, requiring amputation of the affected tissue in very severe cases.

FAINTING

A transient fall in blood pressure causes a momentary reduction in blood flow to the brain and results in fainting (syncope). This may occur when a person stands up suddenly, or following a long period of standing, when blood can pool in the veins of the legs. Lying down and elevating the legs will help restore normal blood pressure.

Areas of blackened skin indicate tissue death caused by prolonged frostbite

LYMPHATIC AND IMMUNE systems

THE BODY HAS A second transport system, which consists of a network of lymphatic vessels and lymphoid organs. Called the lymphatic system, it has two major functions. Firstly, it ensures that blood volume stays the same. Each day, some 24 litres (42 pints) of fluid leave the blood as it passes through the tissues. Most returns to capillaries but some 3–4 litres (5.3–7 pints) remains. This surplus, now called lymph, drains into lymphatic vessels and is emptied back into the bloodstream. Secondly, it plays a major part in body defence. It contains cells, also found in the blood, called lymphocytes and macrophages, which form the immune system, the body's most powerful defence against disease (see pp. 160–1).

LYMPHATIC SYSTEM

There is a one-way flow of lymph along the network of lymphatic vessels. Tiny, blind-ending lymph capillaries pick up lymph in the tissues, and then merge to form larger lymphatic vessels. These eventually drain into two ducts that empty lymph into the subclavian veins, thereby restoring the blood's "lost" fluid. Contracting skeletal muscles surrounding lymphatic vessels help push lymph along them, while valves, like those in veins, prevent backflow. Associated lymphoid organs include the lymph nodes, tonsils, thymus gland, and spleen.

Skin forms a barrier against invading pathogens

Cluster of lymph nodes in armpit

Tonsils trap and destroy eaten or inhaled organisms

Left subclavian vein

Right lymphatic duct receives lymph from the upper right side of the body and empties it into right subclavian vein

Thymus gland processes lymphocytes

Thoracic duct empties into left subclavian vein

Spleen is the largest lymphoid organ and lies next to the stomach

Peyer's patch is a cluster of lymphatic tissue found in the lower part of the small intestine

LYMPHATIC AND IMMUNE SYSTEMS FUNCTIONS

Draining tissue fluid — Lymph vessels form a one-way transport system that drains excess fluid from the tissues and empties it into the blood in order to restore and maintain blood volume.

Transporting dietary fats — Tiny lymph vessels called lacteals collect tiny globules of digested fat from inside the villi of the small intestine, transport them in lymph, and empty them into the blood.

Protection against disease — The immune system consists of cells contained in the lymphatic and cardiovascular systems that protect the body from pathogens and cancer cells.

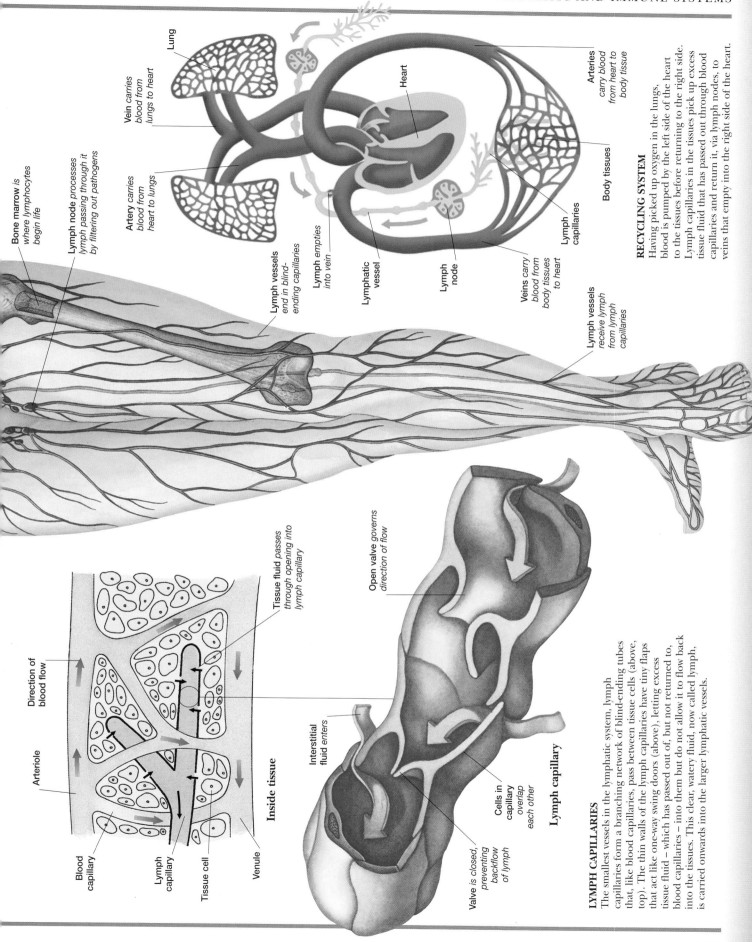

Lung

Vein carries blood from lungs to heart

Heart

Arteries carry blood from heart to body tissue

Bone marrow is where lymphocytes begin life

Lymph node processes lymph passing through it by filtering out pathogens

Artery carries blood from heart to lungs

Body tissues

Lymph vessels end in blind-ending capillaries

Lymph empties into vein

Lymphatic vessel

Lymph node

Veins carry blood from body tissues to heart

Lymph capillaries

Lymph vessels receive lymph from lymph capillaries

RECYCLING SYSTEM

Having picked up oxygen in the lungs, blood is pumped by the left side of the heart to the tissues before returning to the right side. Lymph capillaries in the tissues pick up excess tissue fluid that has passed out through blood capillaries and return it, via lymph nodes, to veins that empty into the right side of the heart.

Open valve governs direction of flow

Tissue fluid passes through opening into lymph capillary

Direction of blood flow

Arteriole

Interstitial fluid enters

Blood capillary

Lymph capillary

Tissue cell

Venule

Inside tissue

Cells in capillary overlap each other

Valve is closed, preventing backflow of lymph

Lymph capillary

LYMPH CAPILLARIES

The smallest vessels in the lymphatic system, lymph capillaries form a branching network of blind-ending tubes that, like blood capillaries, pass between tissue cells (above, top). The thin walls of the lymph capillaries have tiny flaps that act like one-way swing doors (above), letting excess tissue fluid – which has passed out of, but not returned to, blood capillaries – into them but do not allow it to flow back into the tissues. This clear, watery fluid, now called lymph, is carried onwards into the larger lymphatic vessels.

Lymphoid organs

THE LYMPHOID, or lymphatic organs, which include the lymph nodes, spleen, and tonsils, are the parts of the lymphatic system that fight disease. They have a similar structure, being filled with fibres that support macrophages and lymphocytes. Macrophages ingest and destroy pathogens and cancer cells, while lymphocytes play a key role in the immune system (see pp. 160–1), either by attacking pathogens directly, or by disabling them with antibodies. The most numerous lymphoid organs are the lymph nodes, which are scattered throughout the body and filter pathogens out of the lymph passing through them. The spleen removes pathogens from the blood, while the tonsils intercept those passing towards the throat. The thymus gland plays a vital role during childhood in the development of the immune system.

LYMPH NODES

These small, bean-shaped swellings, each 1–25 mm (0.04–1 in) across, occur along lymph vessels like beads on a string. Lymph nodes filter lymph as it passes through them. Surrounded by a tough capsule, the spaces, or sinuses, inside the lymph node are filled with a network of fibres that support macrophages and lymphocytes. These fibres slow the flow of lymph passing through, while macrophages engulf and destroy bacteria, cancer cells, and debris, and lymphocytes launch their immune defences. During infections, lymph nodes may swell up and become tender, a condition known as "swollen glands".

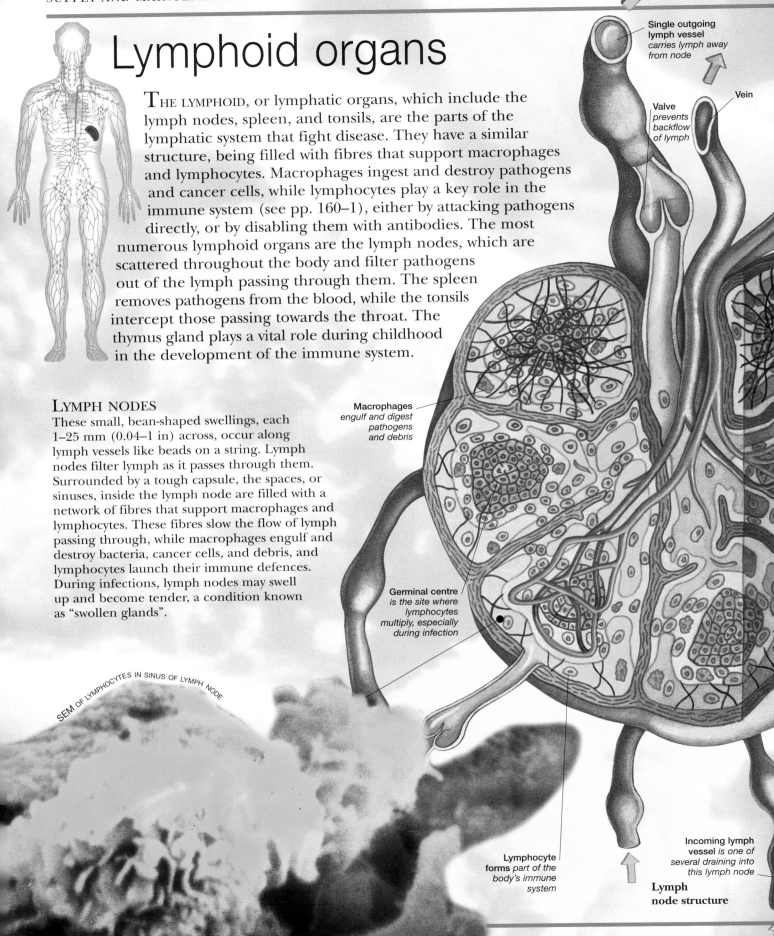

Single outgoing lymph vessel *carries lymph away from node*

Valve *prevents backflow of lymph*

Vein

Macrophages *engulf and digest pathogens and debris*

Germinal centre *is the site where lymphocytes multiply, especially during infection*

SEM OF LYMPHOCYTES IN SINUS OF LYMPH NODE

Lymphocyte forms *part of the body's immune system*

Incoming lymph vessel *is one of several draining into this lymph node*

Lymph node structure

SPLEEN
Largest of the lymphoid organs, the fist-sized spleen lies to the left of the stomach. It receives a rich blood supply through the splenic artery (right) and contains areas of reticular fibres (far right) which support lymphocytes that destroy pathogens. Around these areas are huge numbers of red blood cells, as well as macrophages that process incoming blood by engulfing bacteria, viruses, and worn-out red blood cells.

Artery

Splenic artery
carries blood to spleen

Reticular fibres
form a mesh that supports the cells in the node

Tough capsule
encloses and protects node

Computer graphic of human spleen

SEM OF SECTION OF SPLEEN

Reticular fibres

Sinus *contains fibres that slow the flow of lymph so that macrophages can ingest bacteria and debris*

Thymus gland
plays a role in the body's defences, and is large at birth, but shrinks over several years

Heart

Lung

Uvula

Tonsil

Tonsil

Tongue

TONSILS
A ring of five tonsils – two at the back of the mouth (above), two at the back of the tongue, and one in the upper throat – guard the entrance to the digestive and respiratory systems from bacteria carried in food and air. Bacteria migrate into the tonsils where they are trapped inside deep "crypts" and destroyed by lymphocytes.

THYMUS GLAND
The two-lobed thymus gland carries out its most important functions in the early years of life. It trains lymphocytes to be effective as part of the body's immune system. Unspecialized lymphocytes arrive in the blood from bone marrow. They mature in the thymus, becoming capable of attacking specific pathogens, then leave as T-cells (thymus-dependent) to be posted to lymphoid organs, blood, and lymph.

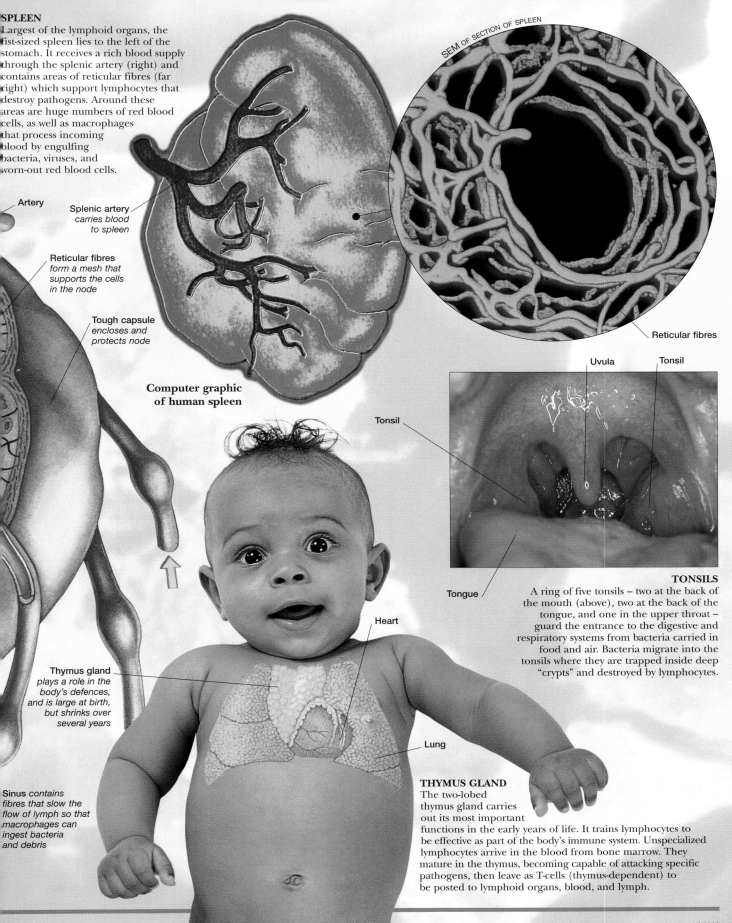

155

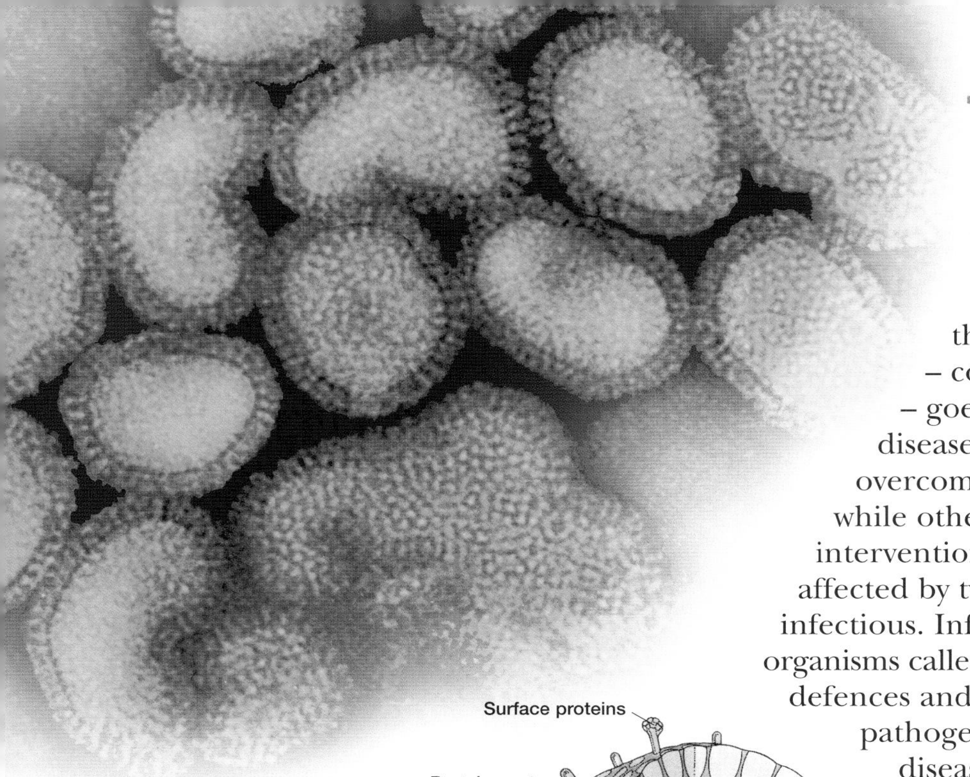

Diseases

From time to time, the body fails to work properly because one or more of the mechanisms that maintain homeostasis – constant, stable conditions inside the body – goes wrong. Any such breakdown is called a disease. Some diseases are short-lived and easily overcome by the body's natural defence systems, while others are more serious and require outside intervention in the form of drugs. Humans are affected by two types of disease – infectious and non-infectious. Infectious diseases are caused by parasitic organisms called pathogens that break through the body's defences and grow and multiply in its tissues. Most pathogens are micro-organisms. Non-infectious diseases are those not caused by pathogens, and include cancers, nervous system disorders and autoimmune conditions (see p. 165).

Viruses

Chemical packages rather than living things, viruses are about one-hundredth the size of bacteria and consist of genetic material, either DNA or RNA, surrounded by a protein coat. Once inside the body, a virus invades a cell, and hijacks its metabolism. It uses this to make copies of itself, which break out and move on to infect other cells. Viral diseases include colds and polio.

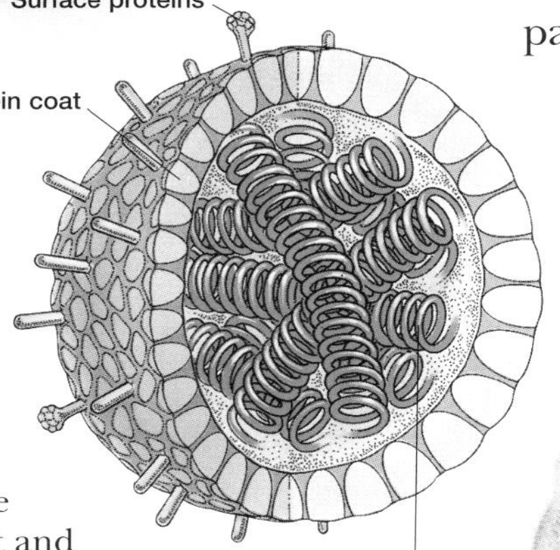

Surface proteins

Protein coat

Cross-section of a virus

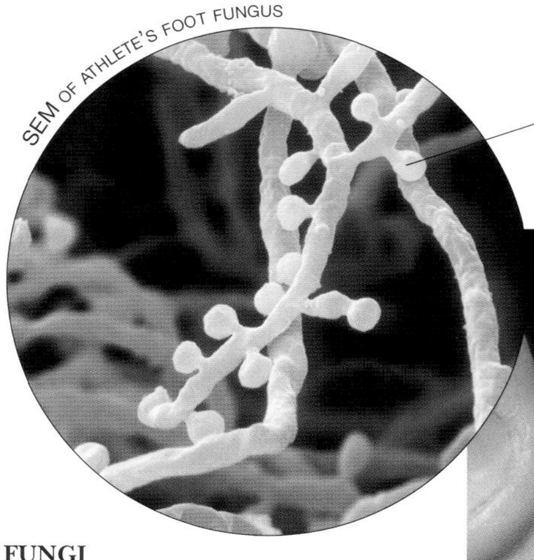

Nucleic acid (DNA or RNA)

Spores *released by the fungus spread infection*

SEM OF ATHLETE'S FOOT FUNGUS

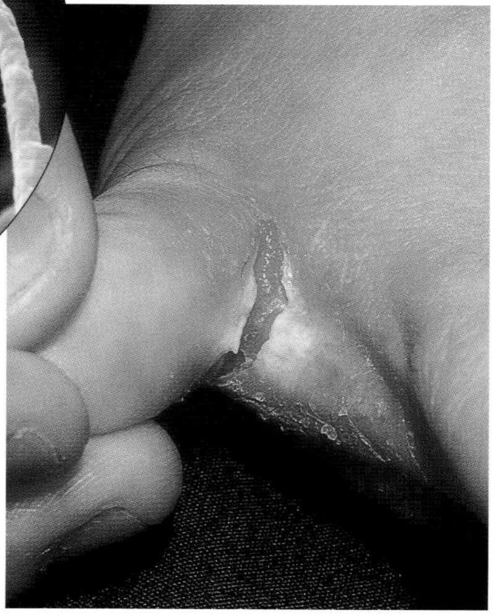

FUNGI

Many fungi are helpful to humans, such as mushrooms or moulds used to produce antibiotics, which feed on dead material. But some fungi are parasitic on humans, including the fungus whose filaments (above) feed on skin and cause athlete's foot (right), and the yeast that causes candidiasis (thrush).

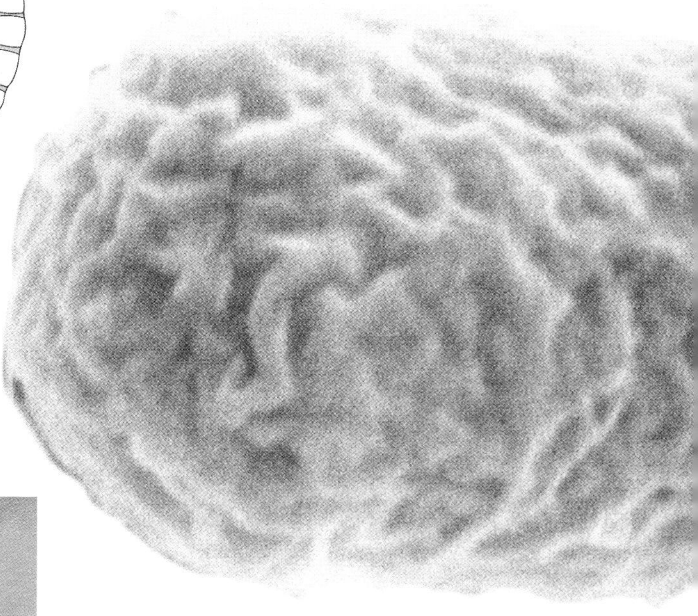

Bacteria

The smallest living organisms, bacteria, are found in their trillions living harmlessly in the soil, air, and in water, as well as in or on the human body. But some bacteria – known more commonly as germs – are pathogenic, and cause diseases such as cholera, diphtheria, whooping cough, tuberculosis (TB), and typhoid. Pathogenic bacteria thrive at body temperature, and reproduce by splitting in two about once every 20 minutes. They damage the body by releasing poisons called toxins.

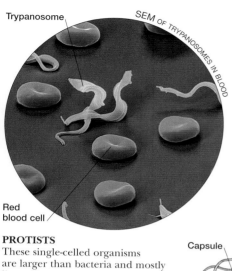

Trypanosome

SEM OF TRYPANOSOMES IN BLOOD

Red blood cell

NON-INFECTIOUS DISEASES

The commonest forms of non-infectious disease include cardiovascular system diseases, such as heart attacks and strokes (see pp. 88–9), respiratory system diseases like bronchitis, and a number of different cancers, in which body cells multiply out of control (right) to form tumours that upset normal body functioning. Causes of cancerous tumours include "faults" in genes passed on from parents, external factors such as chemicals or radiation, and lifestyle factors, such as smoking or lack of exercise.

SEM OF BREAST CANCER CELLS

Cells *have uneven surfaces with protrusions*

PROTISTS

These single-celled organisms are larger than bacteria and mostly live in water or soil. About 30 species of protists are pathogenic, and can cause serious diseases. Some, like the trypanosome (above) that causes sleeping sickness, or the protist that causes malaria, are spread by blood-sucking insects. Others, like *Giardia* (see pp. 190–1), which causes severe diarrhoea, are taken in through contaminated food or water.

Capsule

Flagella

Cell membrane

Pili

Nucleoid region

Cell wall

Cutaway drawing of a bacillus (rod-shaped bacterium)

BACTERIAL STRUCTURE

Bacterial cells are simple compared to those of all other living organisms. They lack membrane-bound organelles such as mitochondria (see pp. 20–1), and their genetic material (DNA) is not enclosed in a nucleus but is located in the nucleoid region. Outside the plasma, or cell, membrane is a protective cell wall and outer capsule. Whip-like flagella move the bacterium, while shorter pili help anchor it to surfaces.

SEM OF ESCHERICHIA COLI BACTERIUM

Escherichia coli produces a toxin that causes cramps and diarrhoea

BACTERIAL TOXINS

By the late 19th century, it had been proved by French scientist Louis Pasteur that bacteria were the cause of many diseases. In 1888, two of his students, Pierre Roux (1853–1933) and Alexandre Yersin (1863–1943), showed how pathogenic bacteria make people ill. Using the bacterium that causes diphtheria, they found that a toxin released by the bacterium causes the disease, not the bacterial cell itself.

Roux in his laboratory

Non-specific defences

THROUGHOUT LIFE, the body is exposed to an array of infectious pathogens that would, if left unchecked, invade, exploit, and ultimately destroy it. Fortunately, the body has a highly sophisticated defence system, which consists of two main parts: the non-specific defences described here, and the immune system (see pp. 160–1). Non-specific defences are built into the body at birth, and respond in the same way to all invading pathogens. First, skin and other outer defences present a physical barrier. Then, if pathogens breach this, a system of defensive cells and anti-microbial chemicals in blood and tissue fluids springs into action.

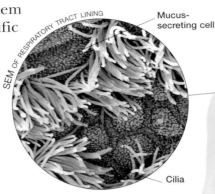
Tear-collecting duct

EYES
Tears produced by glands in the eyelids wash away dirt and contain an antiseptic substance.

Glandular tissue

Enzyme-secreting cells

Mucus-secreting cells

LM OF SECTION THROUGH A TEAR GLAND

SEM OF RESPIRATORY TRACT LINING

Mucus-secreting cell

Cilia

RESPIRATORY TRACT
Mucus is produced in the lining of the respiratory tract. It traps pathogens and is carried to the throat by cilia.

LM OF SECTION THROUGH

MOUTH
Watery saliva, produced by glands around the mouth, washes the mouth out, and contains bacteria-killing lysozyme.

TEM OF SECTION THROUGH INTESTINAL LINING

Mucus

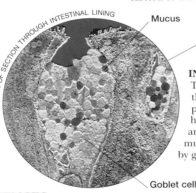

INTESTINES
The lining of the intestines is protected from harmful organisms and chemicals by mucus produced by goblet cells.

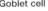
Goblet cell

Opening of gastric gland

Stomach lining

SEM OF GASTRIC GLAND

STOMACH
Glands in the stomach lining produce hydrochloric acid which kills most invading organisms.

OUTER DEFENCES

This diagram shows the main parts of the body's outer defences. The tightly knit cells of the skin, and those of the mucous membrane lining the respiratory, digestive, urinary, and reproductive systems, help stop pathogens entering the tissues. Tears contain a bacteria-killing substance called lysozyme, as do saliva and sweat. Mucus lining the respiratory system traps pathogens, then passes them to the throat for swallowing. Stomach acid destroys most swallowed bacteria. On the skin, and in the digestive and female reproductive system, colonies of harmless bacteria prevent harmful bacteria from establishing themselves.

LM OF BACTERIA IN GENITAL TRACT

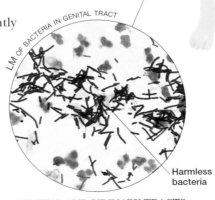

Harmless bacteria

GENITAL AND URINARY TRACTS
The expulsion of urine in the urinary tract and the presence of harmless bacteria in the genital tract prevent harmful organisms from multiplying in these areas.

Hair

Skin surface

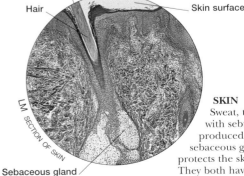

LM SECTION OF SKIN

Sebaceous gland

SKIN
Sweat, together with sebum produced by sebaceous glands, protects the skin. They both have mildly antiseptic properties.

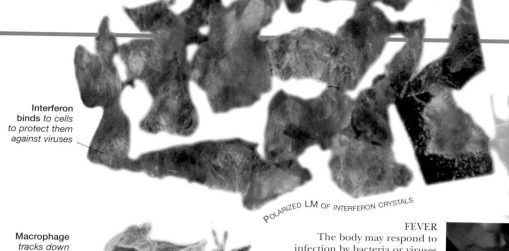

Interferon binds *to cells to protect them against viruses*

POLARIZED LM OF INTERFERON CRYSTALS

ANTIMICROBIAL SUBSTANCES

Two sets of blood proteins – interferon and the complement system – form a key part of the defence force. Interferon is released by body cells already infected with viruses. It stimulates neighbouring cells to protect themselves against viral infection. The complement system of over 20 proteins aids the inflammation process (see below). The proteins attach themselves to bacteria either to make them more "tasty" for phagocytes to eat, or to destroy them by making the cell membranes burst open.

FEVER

The body may respond to infection by bacteria or viruses by raising its temperature above the normal 37°C (98.6°F) in order to stop the invaders multiplying. This strategy, called fever or pyrexia, is often accompanied by sweating, shivering, and a feeling of thirst. It is triggered by white blood cells that release chemicals called pyrogens which reset the body's "thermostat" in the brain's hypothalamus so that body temperature rises above normal.

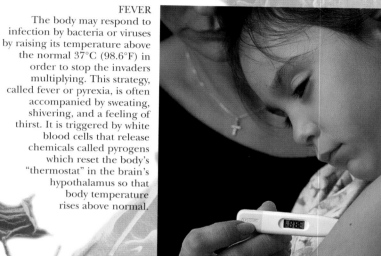

Macrophage *tracks down pathogen*

Extension of macrophage *engulfs protist prior to digestion*

Parasitic protist *that causes the tropical disease leishmaniasis*

SEM OF MACROPHAGE ENGULFING PROTIST

CELL EATERS

Phagocytes are white blood cells that engulf and destroy invading organisms. There are two types: neutrophils and macrophages. Neutrophils circulate in the blood before being transferred to the tissues, where they seek out organisms. Macrophages also start in the bloodstream, where they are known as monocytes, before moving to the tissues, where they become macrophages. Some macrophages stay in one place; others travel around looking for infective organisms. All phagocytes flow around organisms, wrapping them within a membrane that fuses with granules called lysosomes. These granules contain strong chemicals that digest the organisms, producing harmless substances that pass out through the cells' outer membrane.

Chemicals *that attract white blood cells*

Pathogen

Injured skin

White blood cells *engulf bacteria*

Inflamed tissue

Blood vessel *widens*

Damage

Response

INFLAMMATORY RESPONSE

This is the familiar warm, reddish, tender swelling that appears after an injury. At the site of the damage, tissue cells release histamine and other chemicals. They make the blood vessels wider, so that extra blood arrives, and more leaky, so that fluid passes into the tissues, aiding repair and making the area red, swollen, and warm. These chemicals also attract phagocytes that destroy pathogens.

Immune system

THE MOST POWERFUL PART of the body's defences, the immune system consists of billions of white blood cells, called lymphocytes, found in the circulatory and lymphatic systems, and in other tissues. While non-specific defences provide unchanging protection against all pathogens, the immune system attacks specific pathogens and remembers them so that if they should attack again it can respond with lightning speed. This gives a person long-term protection, or immunity, against diseases. The immune system is triggered into action by foreign antigens, or cell markers, that identify pathogens and cancer cells, but not their own body cells. It has two linked parts. The humoral part works by releasing antibodies to disable pathogens, while the cellular part directly attacks and destroys invaders.

Antigen of bacterium

Macrophage

Bacterium
enters the body and is surrounded by a macrophage cell

Non-matching B-cells

Engulfed bacterium

Antigen of bacterium

Matching B-cell
multiplies to produce plasma cells and memory B-cells – plasma cells create antibodies to destroy bacteria, while memory B-cells are stored for future use

Memory B-cell
rapidly produces plasma cells when it encounters the same bacteria second-time around

Inactivated bacterium

Plasma cell

Antibodies
released by the plasma cells lock on to bacterial antigens and inactivate the bacterium; the antibodies also attract phagocytes to the area to help destroy the bacteria

Antigen of bacterium

Phagocyte
destroys bacterium

ANTIBODIES

Also called immunoglobulins, these proteins are made by B-lymphocytes and are found in blood, lymph, and other tissue fluids. Most of each Y-shaped antibody molecule is identical in all antibodies, but part of the "arms" of the Y are unique to each particular type of antibody. It is this unique, variable region that binds the antibody to a specific antigen carried by its target pathogen, just as a key fits into a lock. Disabled by antibodies, the pathogen is now marked for destruction.

COMPUTER MODEL OF AN ANTIBODY

HUMORAL IMMUNITY

This part of the immune system uses antibodies to attack invaders – particularly bacteria and some viruses – and it involves lymphocytes called B-cells which are found in the lymphoid organs of the lymphatic system (see pp. 154–55). A pathogen entering the body is engulfed by a pathogen-destroying cell called a macrophage. This presents the antigens carried by the pathogen to a matching B-cell which is primed to recognize that specific antigen. The B-cell then divides rapidly to produce plasma cells which flood blood, lymph, and tissue fluid with antibodies, and memory B-cells which remember that pathogen and will respond rapidly to it if it returns. The antibodies bind to the pathogens and mark them for destruction by phagocytes or chemicals.

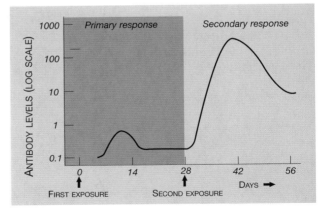

IMMUNE RESPONSE
When a pathogen invades the body for the first time, its antigens stimulate the immune system to produce antibodies. This primary response takes several days (see above). If, subsequently, a second invasion takes place, the immune system leaps into action, releasing high levels of antibodies that target the specific pathogen. A person is now immune to the disease caused by that pathogen, and should never suffer from it again.

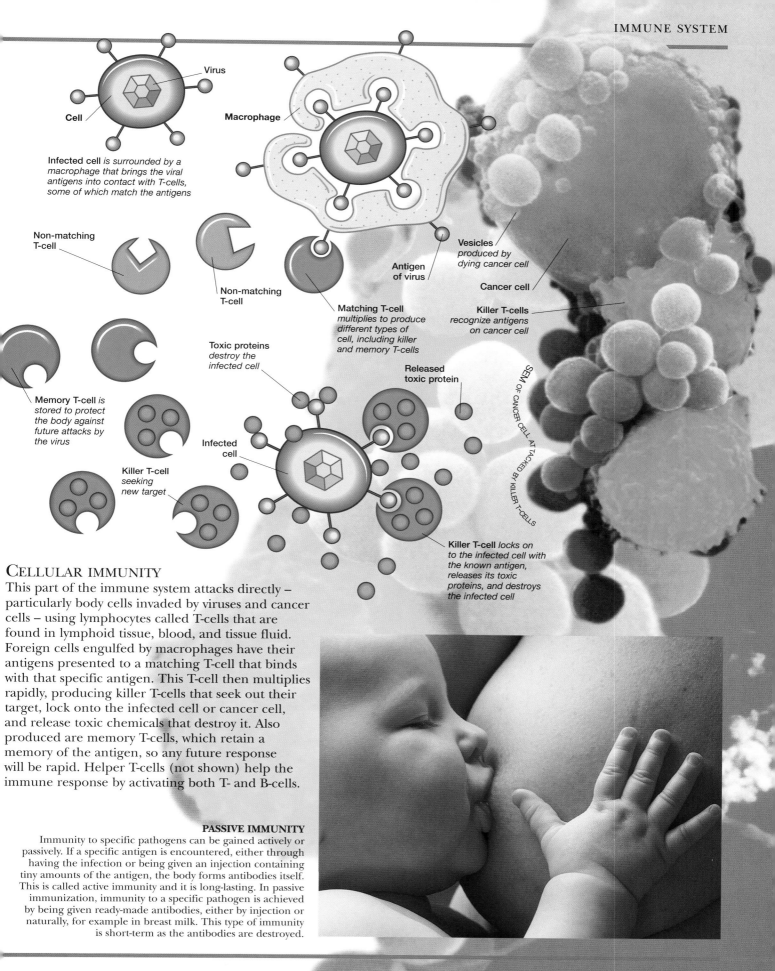

Infected cell *is surrounded by a macrophage that brings the viral antigens into contact with T-cells, some of which match the antigens*

Cell

Virus

Macrophage

Non-matching T-cell

Non-matching T-cell

Antigen of virus

Matching T-cell *multiplies to produce different types of cell, including killer and memory T-cells*

Vesicles *produced by dying cancer cell*

Cancer cell

Killer T-cells *recognize antigens on cancer cell*

Toxic proteins *destroy the infected cell*

Released toxic protein

Memory T-cell *is stored to protect the body against future attacks by the virus*

Killer T-cell *seeking new target*

Infected cell

Killer T-cell *locks on to the infected cell with the known antigen, releases its toxic proteins, and destroys the infected cell*

SEM OF CANCER CELL ATTACKED BY KILLER T-CELLS

CELLULAR IMMUNITY

This part of the immune system attacks directly – particularly body cells invaded by viruses and cancer cells – using lymphocytes called T-cells that are found in lymphoid tissue, blood, and tissue fluid. Foreign cells engulfed by macrophages have their antigens presented to a matching T-cell that binds with that specific antigen. This T-cell then multiplies rapidly, producing killer T-cells that seek out their target, lock onto the infected cell or cancer cell, and release toxic chemicals that destroy it. Also produced are memory T-cells, which retain a memory of the antigen, so any future response will be rapid. Helper T-cells (not shown) help the immune response by activating both T- and B-cells.

PASSIVE IMMUNITY

Immunity to specific pathogens can be gained actively or passively. If a specific antigen is encountered, either through having the infection or being given an injection containing tiny amounts of the antigen, the body forms antibodies itself. This is called active immunity and it is long-lasting. In passive immunization, immunity to a specific pathogen is achieved by being given ready-made antibodies, either by injection or naturally, for example in breast milk. This type of immunity is short-term as the antibodies are destroyed.

VACCINATION FOR ALL

THROUGHOUT HISTORY, people have lived in fear of infectious diseases, including smallpox, which killed 40 per cent of those infected and left gruesome scars on the faces of survivors. By the 17th century, smallpox had become the most serious infectious disease in the West. The turning point in the fight against it came with the discovery of vaccination, when exposure to harmless cowpox was found to protect people from catching its deadlier relative. The technique, pioneered by English doctor Edward Jenner in the late 18th century, has saved millions of lives, but it was preceded by another, albeit more risky, method of prevention.

VARIOLATION

In 1717, a technique called variolation – which originated in 10th-century China – came to the notice of Lady Mary Wortley Montagu (1689–1762), wife of the British Ambassador to Turkey. Turkish children were inoculated by scratching their skin and applying pus taken from the blisters of people with mild smallpox. Lady Mary, who had her own children treated in this way, introduced variolation when she returned to England in 1721. But there was an element of chance in the procedure, with some patients developing full-blown smallpox.

TURKISH TECHNIQUE
Lady Mary Wortley Montagu noticed that Turkish women carried pus from a mild form of smallpox in walnut shells, and used it to inoculate children to protect them from the dangerous form of the disease. She encouraged people in England to use the technique, called variolation.

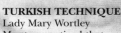
Ivory blade

Tortoiseshell handle

Vaccination lancet

Large, pus-filled sore

Drawing from Jenner's book showing cowpox

LIFE-SAVER
Within three years of Edward Jenner's discovery, more than 100,000 people had been vaccinated against smallpox. His technique became the most important life-saving discovery of the 18th century, and was the forerunner of today's mass immunization programmes against infectious diseases.

Sculpture showing Jenner vaccinating James Phipps

VACCINATION
Jenner used a lancet (above, left) to scratch James Phipp's arm, then inoculated him with pus taken from a blister on a milkmaid's hand. Phipps became immune to smallpox – a disease that, in Jenner's day, killed 3,000 people a year in London alone.

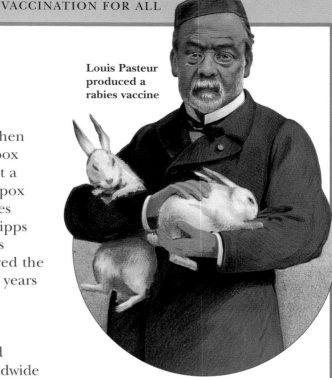

Louis Pasteur produced a rabies vaccine

JENNER'S BREAKTROUGH

It was Edward Jenner (1749–1823) who developed a more reliable method of preventing smallpox. He noticed that milkmaids contracted a similar but milder infection called cowpox from their cows but never caught smallpox, even when in close contact with smallpox victims. Reasoning that cowpox infection could protect against smallpox, Jenner carried out a bold experiment. On 14 May 1796, he took pus from a cowpox blister and introduced it into the arm of eight-year-old James Phipps. Six weeks later, when he deliberately inoculated Phipps with smallpox, the boy did not develop the disease. Jenner's treatment, called vaccination, quickly caught on, and inspired the great French bacteriologist Louis Pasteur, who nearly 100 years later produced a vaccine to protect against rabies.

THE END OF SMALLPOX

Even in the mid-1960s, smallpox still infected 10 million and killed 2 million people worldwide every year. But a concerted vaccination programme by the World Health Organization (WHO) finally eliminated the disease in 1977. This was the first time in history that a disease had been eradicated by human intervention. It now only exists in laboratories in Russia and the United States.

PROTECTING CHILDREN

Although smallpox has been defeated, many other infectious diseases still kill millions. Vaccinating, or immunizing, children remains the single most important preventative treatment. Vaccines – medications that prime the immune system to fight infection – have been developed to protect against many life-threatening diseases. In several countries, health authorities implement an immunization schedule for children which starts at two months and continues until the teenage years.

EARLY PROTECTION
Babies are very vulnerable to infection, so immunization provides vital protection.

FIGHTING RABIES
In 1885, Louis Pasteur (1822–95) tested a vaccine, made from rabbit spinal cord, on Joseph Meister, a nine-year-old who had been bitten by rabid dogs. Meister recovered from rabies, a horrific and fatal disease, and public acclaim for Pasteur's achievement led to the foundation of the Pasteur Institute in 1888.

HOW IMMUNIZATION WORKS

Immunization, or vaccination, prepares the body to fight a specific infection. In active immunization, shown here, a vaccine is injected into a person to stimulate their immune system to produce antibodies (see pp. 160–61) that will destroy the "real" pathogen if it invades the body.

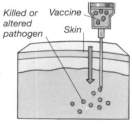

Killed or altered pathogen *Vaccine* *Skin*

1. To immunize them against a specific disease, a person is injected with a vaccine containing a killed or altered version of the pathogen that causes the disease.

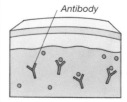

Antibody

2. Within days, the immune system produces antibodies against the disease, as well as long-lived memory cells that "remember" the pathogen's identity for future reference.

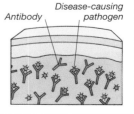

Antibody *Disease-causing pathogen*

3. Should the "real" pathogen enter the body at a later date, the immune system launches an immediate response, producing large numbers of antibodies that immobilize the invader.

Immune system disorders

THE IMMUNE SYSTEM of the human body can be disrupted. Allergies, such as hay fever, are common and result from an excessive immune reaction to what would normally be regarded as a harmless substance. Extreme allergies can cause anaphylaxis, a rare but life-threatening condition. Sometimes the immune system can react inappropriately, rather than excessively, by forming antibodies against its own cells. This is known as autoimmunity. Even though they are appropriate, immune responses to transplanted tissue and organs are unwanted, so drugs are taken long-term to try to prevent the immune system from rejecting the new transplants. Immunodeficiency (weakening of the immune system) may develop for a number of reasons, including diseases of the bone marrow, kidney failure, and the human immunodeficiency virus (HIV).

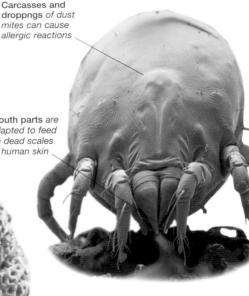

TESTING FOR ALLERGIES
Skin-prick tests are often performed on individuals with allergies to try to identify the allergens responsible. Small drops of allergen solutions are placed on the skin, which is then pricked with a needle. If the test is positive, a red lump will form at the site of the allergen, usually within half an hour.

ALLERGIES

When the immune system reacts excessively to specific antigens, allergies are said to be present. Various substances, known as allergens, can provoke such a response. Common allergens include the house-dust mite and pollen. Allergens may have their effect by being breathed in, by being eaten, or through skin contact. When the body encounters an allergen for the first time, it becomes "sensitized". On subsequent encounters the immune system mounts an excessive response. Allergies may cause a variety of conditions. Hay fever, one of the most common, is a reaction to pollen and produces symptoms of watery eyes, runny nose, and sneezing.

SEM OF POLLEN GRAINS

POLLEN
These minute grains are released by plants as part of their reproductive process. Although some plants use insects to transfer their pollen, others release enormous quantities of it into the air, where humans may breathe it in.

Carcasses and droppngs *of dust mites can cause allergic reactions*

Mouth parts *are adapted to feed on dead scales of human skin*

DUST MITES
Up to 5,000 of these tiny relatives of spiders can live in a single gram of house dust (140,000 in an ounce). They feed on human skin flakes, and release microscopic droppings that can cause allergies when breathed in.

Wind-borne pollen *can cause allergies and hay fever*

REJECTED TISSUES

The first attempts at transplanting human tissue from one person to another were made early in the 20th century. They were unsuccessful because the body "rejected" the transplanted tissue. In the late 1940s, British immunologist Peter Medawar (1915–87) showed that the immune system regards the cells of the transplanted tissue as foreign – as it would pathogenic bacteria – and destroys it. His work helped to make organ transplantation – in which drugs are given to suppress the immune system – possible. In 1960, Medawar was awarded the Nobel Prize for Medicine.

Common wasp *stinging a human*

ANAPHYLACTIC SHOCK

Some people have excessive and potentially life-threatening allergic responses to specific allergens, such as wasp stings and peanuts. The response is so strong that contact with the allergen results in a major fall in blood pressure. Symptoms of a severe allergic response (anaphylactic shock) may include difficulty in breathing, a skin rash, and loss of consciousness. The condition, which requires urgent treatment, is fortunately rare.

Vitiligo causes the skin to lose its pigment

Whitened area *shows pigment loss*

AUTOIMMUNITY

If the body does not recognize its own cells, it will regard them as foreign and produce an immune response, which attacks and damages them. Known as autoimmunity, this forms the basis for a number of conditions affecting various parts of the body. An example is rheumatoid arthritis, in which antibodies are produced against the lining of joints, causing pain and swelling. Another autoimmune disorder is vitiligo, in which antibodies cause patches of skin to lose their colour. It is still not known why autoimmunity develops, but genetic factors are thought to play a part in many of these diseases.

HIV AND AIDS

There are several reasons why the immune system can fail to work properly: one of them is the human immunodeficiency virus (HIV). This virus infects T-helper cells. These are lymphocytes (see pp. 160–1) that play a key role in enabling the immune system to target and destroy pathogens and cancer cells. If the number of T-helper cells drops significantly, serious infections and certain types of cancer can develop due to major weakening of the immune system. This is called acquired immunodeficiency syndrome (AIDS). Drugs have now been developed that can limit the progress of the virus.

HIV particle

Infected T-cells *typically have a lumpy appearance*

SEM OF T-CELL INFECTED WITH HIV

RESPIRATORY system

WHILE THE BODY can do without food or water for a short time, it cannot survive without a continuous supply of oxygen. Its trillions of cells relentlessly consume oxygen in order to release from sugars the energy to power their activities This process – called cell, or internal, respiration – also produces waste carbon dioxide. The body's oxygen supply is provided by the respiratory system, which draws air into the body, transfers its oxygen to the bloodstream, then pushes air out, expelling unwanted carbon dioxide.

SEM OF CILIA INSIDE NASAL CAVITY

Air passes through the nasal cavity, being warmed, moistened, and cleaned as it does so

Hairs in nostrils filter out large particles

NASAL CAVITY
The entrance to the respiratory system is through the nasal cavity, the hollow space behind the nose. Air contains dust and dirt particles that could damage the lungs if they got that far. Fortunately, the nose and nasal cavity provide a filtration service. Hairs guarding the nostrils remove larger particles as air is inhaled (breathed in). Then dust is trapped by sticky mucus, secreted by the nasal cavity lining, which is then moved by cilia to the back of the throat for swallowing. Inhaled air is moistened and warmed as it passes through the nasal cavity.

LUNGS AND AIRWAYS
The respiratory system consists of the lungs and the airways, or tubes – nose, pharynx (throat), larynx (voice box), trachea (windpipe), and bronchi – that carry air between the lungs and the outside atmosphere. Inhaled air travels along the nose, pharynx, larynx, and trachea before entering one of two branches – the bronchi – that enter both lungs. Inside the lungs, bronchi divide into smaller and smaller branches that finally end in sac-like alveoli where oxygen and carbon dioxide are exchanged. Exhaled air returns in the opposite direction.

ENDOSCOPIC VIEW OF TRACHEA

SEM OF LINING OF TRACHEA

TRACHEA
The trachea (windpipe) is the flexible tube through which air travels from the larynx towards the lungs. At its lower end, it divides into two main bronchi, one for each lung. The trachea is held open by C-shaped rings of cartilage embedded in its walls. The lining of the trachea continues the work of the nasal cavity in removing dust and dirt from air.

Right lung

Cilia projecting from cells lining the trachea move rhythmically to waft dust-laden mucus up to the throat so it can be swallowed or spat out

RESPIRATORY SYSTEM FUNCTIONS

Ventilation	*Muscle contractions alter the volume of the chest, drawing air along the respiratory tract and into and out of the lungs.*
External respiration	*Within the lungs, oxygen diffuses from the air into the bloodstream, and carbon dioxide diffuses in the opposite direction.*
Internal respiration	*Throughout the body, oxygen diffuses out of the blood into cells, where it is used in the chemical processes that release energy. Carbon dioxide diffuses in the opposite direction.*

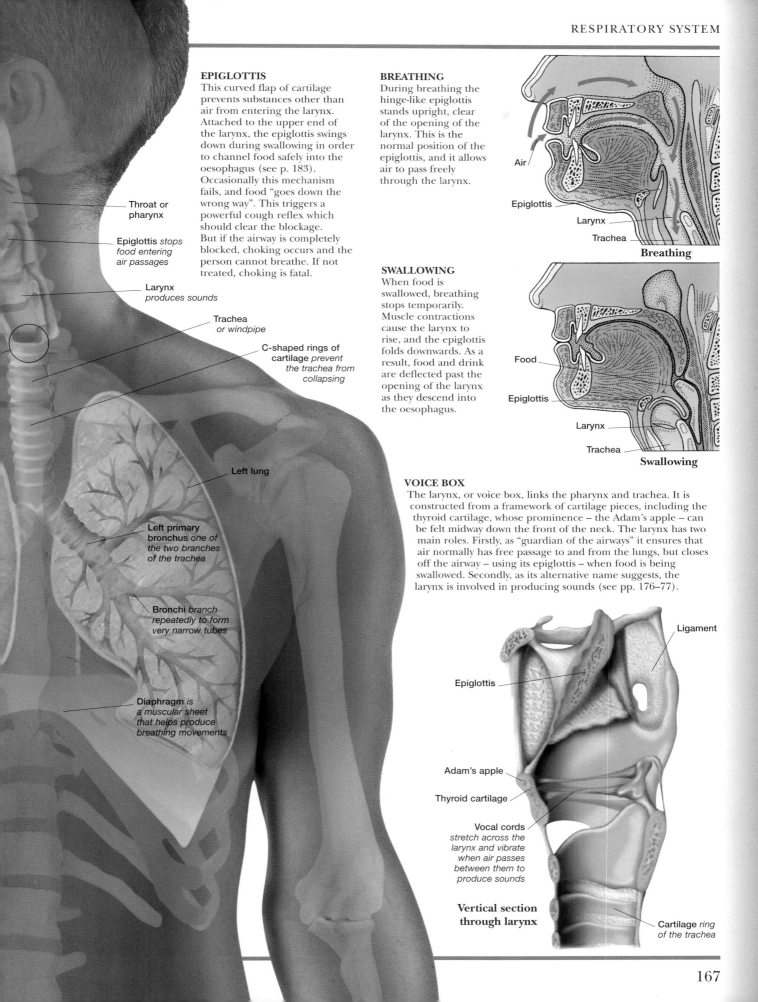

EPIGLOTTIS

This curved flap of cartilage prevents substances other than air from entering the larynx. Attached to the upper end of the larynx, the epiglottis swings down during swallowing in order to channel food safely into the oesophagus (see p. 183). Occasionally this mechanism fails, and food "goes down the wrong way". This triggers a powerful cough reflex which should clear the blockage. But if the airway is completely blocked, choking occurs and the person cannot breathe. If not treated, choking is fatal.

Throat or pharynx

Epiglottis *stops food entering air passages*

Larynx *produces sounds*

Trachea *or windpipe*

C-shaped rings of cartilage *prevent the trachea from collapsing*

Left lung

Left primary bronchus *one of the two branches of the trachea*

Bronchi *branch repeatedly to form very narrow tubes*

Diaphragm *is a muscular sheet that helps produce breathing movements*

BREATHING

During breathing the hinge-like epiglottis stands upright, clear of the opening of the larynx. This is the normal position of the epiglottis, and it allows air to pass freely through the larynx.

Air

Epiglottis

Larynx

Trachea

Breathing

SWALLOWING

When food is swallowed, breathing stops temporarily. Muscle contractions cause the larynx to rise, and the epiglottis folds downwards. As a result, food and drink are deflected past the opening of the larynx as they descend into the oesophagus.

Food

Epiglottis

Larynx

Trachea

Swallowing

VOICE BOX

The larynx, or voice box, links the pharynx and trachea. It is constructed from a framework of cartilage pieces, including the thyroid cartilage, whose prominence – the Adam's apple – can be felt midway down the front of the neck. The larynx has two main roles. Firstly, as "guardian of the airways" it ensures that air normally has free passage to and from the lungs, but closes off the airway – using its being epiglottis – when food is being swallowed. Secondly, as its alternative name suggests, the larynx is involved in producing sounds (see pp. 176–77).

Ligament

Epiglottis

Adam's apple

Thyroid cartilage

Vocal cords *stretch across the larynx and vibrate when air passes between them to produce sounds*

Vertical section through larynx

Cartilage *ring of the trachea*

Lungs

THESE TWO ORGANS SURROUND the heart, occupying most of the space inside the thorax. While the rest of the respiratory system is concerned with getting air into and out of the body, the lungs concentrate on getting oxygen into, and carbon dioxide out of, the bloodstream. To do this, they interact closely with the circulatory system, whose mass of blood vessels gives the lungs their pinky-red colour. The lungs have a spongy feel as a result of their internal structure – a system of air-filled, progressively branching tubes terminating in microscopic "air bags" through which oxygen enters the blood and waste carbon dioxide leaves it.

CT SCAN OF SECTION THROUGH THORAX

Heart

Lung

The right lung is separated into three lobes, the left lung into two

LUNG STRUCTURE

The lungs are light, spongy structures that are approximately conical in shape. The uppermost part of each lung extends above the clavicle (collar bone) into the neck, and their bases rest on the diaphragm. Each lung is divided into separate portions called lobes: the right lung consists of three lobes, whereas the left lung, which is slightly smaller in order to make space for the heart, consists of two. Surrounding the lungs are two membranes called pleura, between which lies a thin layer of pleural fluid which ensures that the lungs expand and shrink smoothly with each breath. The ribcage protects the lungs, and its muscles assist in breathing.

BRONCHIAL TREE

This resin cast (right) shows the system of airways that carries air into the lungs. The trachea divides into two primary, or main, bronchi, each supplying one lung. These split into secondary bronchi, which then subdivide into narrower, tertiary bronchi. Bronchi further divide into terminal bronchioles. This structure is called the bronchial tree as it resembles an upside-down tree with the trachea as "trunk", bronchi as "branches", and bronchioles as "twigs".

Trachea

Primary (or main) bronchus

Respiratory bronchiole

Bronchiole

ALVEOLAR SACS

As terminal bronchioles penetrate more deeply into the lungs, they divide into microscopic respiratory bronchioles. These lead into blind-ending alveolar sacs resembling bunches of grapes. Each "grape", or alveolus, shares with other alveoli an opening into a duct connecting it to the respiratory bronchiole. Alveoli are the site of gas exchange.

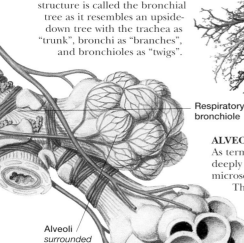

Alveoli *surrounded by blood capillaries*

Tertiary bronchus *branches from the secondary bronchus, which subdivides repeatedly to form terminal bronchioles*

Secondary bronchus *– there are three secondary bronchi in the right lung, each supplying one lobe of the lung*

Right primary bronchus

Trachea

Terminal bronchioles
– there are about
30,000 of these tiny
bronchioles in each lung;
they divide to form
respiratory bronchioles
that lead into the alveoli

SEM OF MACROPHAGES IN LUNG

Left primary
bronchus

DUST EATERS
Breathed-in air carries foreign particles such as dust, pollen, and bacteria, but most of these are trapped in sticky mucus in the nose, trachea, and bronchi. Any tiny particles that do reach the alveoli are engulfed by wandering cells called alveolar macrophages, or dust cells, before they can accumulate and interfere with gas exchange. This SEM shows a macrophage elongating to engulf a dust particle (green). Dust cells are less active in the lungs of cigarette smokers.

ASTHMA
This breathing problem occurs when smooth muscle fibres surrounding the lungs' bronchioles contract, and bronchiole linings become inflamed and swollen. This causes narrowing of the airway, resulting in episodes of breathlessness, wheezing, and coughing. This asthma sufferer is using an inhaler to breathe in a bronchodilator drug that relaxes the muscle fibres and widens the bronchioles. Asthma can be triggered by many factors including pollen, house dust, or air pollution.

LUNG CANCER
These two photographs show clearly the difference between a healthy lung and one affected by cancer. Lung cancer is the most common form of cancer in Western countries. However it is easily prevented, as almost all cases are caused by cigarette smoking. Symptoms such as chest pain, coughing, and weight loss may occur as the cancer develops within the lung, but in some cases no symptoms appear until the disease is advanced. Cancerous cells tend to spread rapidly from the lung to other areas such as the brain and bones, causing further harm to the patient.

Healthy lung

Cancerous lung

Diaphragm
is a dome-shaped muscle
which separates the
thorax from the abdomen
and plays a key role in
breathing

Pleural membranes
are two membrane layers
covering each lung, which
are separated by
lubricating fluid that
allows them to slide over
each other during
breathing

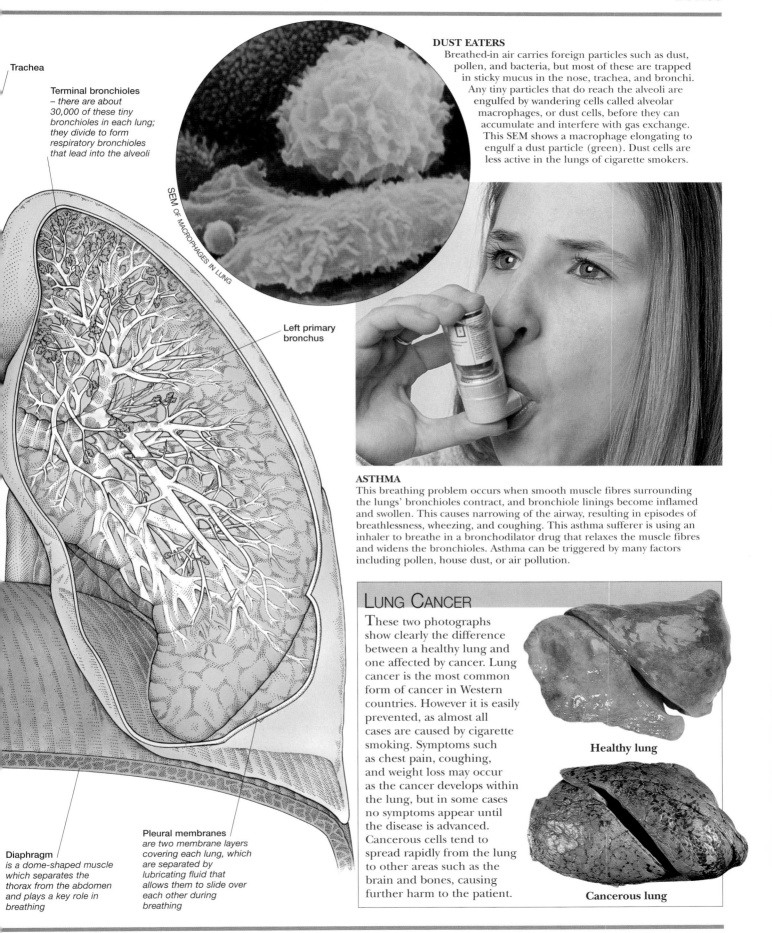

Gas exchange

EVERY MINUTE, LARGE AMOUNTS OF life-giving oxygen is taken into the bloodstream, while potentially poisonous carbon dioxide is expelled. This happens through a mechanism called gas exchange, which occurs in the lungs' tiny sac-like alveoli. Two features of alveoli make gas exchange fast and efficient. Firstly, the wall of an alveolus is just one cell thick, as is the wall of the blood capillaries that surround it. Where the two meet, they form a respiratory membrane just 0.0005 mm (0.00002 in) wide, across which oxygen can move rapidly into, and carbon dioxide out of, the blood. Secondly, the two lungs contain some 300 million alveoli that collectively provide a surface area for gas exchange of 70 sq m (750 sq ft) – 35 times the surface area of the skin – squeezed into a space inside the chest that is no bigger than a shopping bag.

Before diffusion   After diffusion

DIFFUSION
The natural tendency of molecules to move randomly from an area of high concentration to one of low until evenly spread out is called diffusion. Some molecules (above, red) can diffuse through cell membranes (green). This is exactly what happens to oxygen during gas exchange in the alveoli, except that breathing brings more oxygen into the alveoli, while blood capillaries carry it away, never allowing it to be evenly distributed.

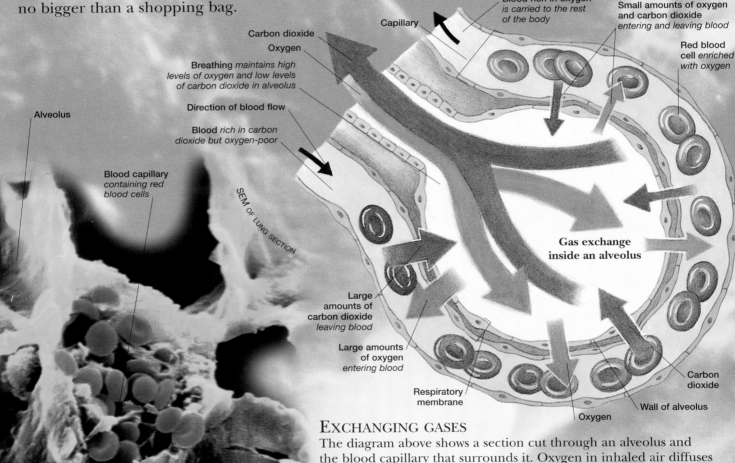

Capillary

Carbon dioxide

Oxygen

Breathing *maintains high levels of oxygen and low levels of carbon dioxide in alveolus*

Direction of blood flow

Blood *rich in carbon dioxide but oxygen-poor*

Alveolus

Blood capillary *containing red blood cells*

SEM OF LUNG SECTION

Blood rich in oxygen *is carried to the rest of the body*

Small amounts of oxygen and carbon dioxide *entering and leaving blood*

Red blood cell *enriched with oxygen*

Gas exchange inside an alveolus

Large amounts of carbon dioxide *leaving blood*

Large amounts of oxygen *entering blood*

Respiratory membrane

Oxygen

Wall of alveolus

Carbon dioxide

EXCHANGING GASES
The diagram above shows a section cut through an alveolus and the blood capillary that surrounds it. Oxygen in inhaled air diffuses across the thin respiratory membrane, and passes into red blood cells. Inhalation replenishes oxygen supplies in the alveolus, while blood flow removes oxygen-enriched blood. This creates a diffusion "gradient" across the respiratory membrane – high levels of oxygen in the alveolus, low levels in the blood – that ensures a constant flow of oxygen into the bloodstream. The same applies to carbon dioxide, but in the opposite direction. Carbon dioxide diffuses from newly arrived blood into the alveolus, where levels of carbon dioxide are low because it is continually exhaled.

DISCOVERING OXYGEN

During Harvey's lifetime, there was considerable interest in why both an animal and a flame need air to survive. This question was explored by English physicist Robert Boyle (1627–91). The invention of the vacuum pump, which sucked air out of sealed containers, allowed Boyle to show that animals could not survive without air, nor would a candle burn. English doctor John Mayow (1640–79) showed that if a mouse, or a lighted candle, was placed in a confined space, only part of the air was used up before the animal died, or the candle went out, a part that he called nitroaerial spirit. It was left to French chemist Antoine Lavoisier (1743–94) to identify this as oxygen. Lavoisier showed that the human body consumes oxygen during cell respiration and exhales carbon dioxide, just as a burning candle consumes oxygen and generates carbon dioxide. He thought cell respiration occurred only in the lungs, but Italian physiologist Lazzaro Spallanzani (1729–99) proved that cell respiration occurs in every tissue of the body.

ANTOINE LAVOISIER
Seen here experimenting in his laboratory, Antoine Lavoisier gave the name "oxygen" to the part of air that supports life. Lavoisier lost his own life prematurely when he was guillotined by French revolutionaries in 1794.

from food. These reactions, which occur inside cell organelles called mitochondria, release energy to power cell activities. They are also the source of the heat thought by Aristotle to be generated by the vital flame.

RELEASING ENERGY

By 1856, it was known that muscles respire, taking in oxygen, giving out heat, and generating energy for their contraction. It was not until the 1930s, however, that German-born British biochemist Hans Krebs (1900–81) described the complex chemical reactions of aerobic respiration (called the Krebs cycle) that use oxygen – produced by the world's plant life – to break down glucose derived

Cell respiration takes place on these inner fields

TEM OF A MITOCHONDRION

Oxygen *is released by leaves as a product of photosynthesis*

Leaves *absorb carbon dioxide from the air*

OXYGEN PRODUCERS
Plants generate oxygen, which is essential for life. They take in water through their roots and absorb carbon dioxide from air into their leaves. From these, in a reaction called photosynthesis, they manufacture sugars used for growth and release oxygen as a waste product.

MITOCHONDRIA
These sausage-shaped organelles are found inside body cells. Oxygen, taken in through the lungs and carried to cells by the blood, is used inside mitochondria to help release energy from glucose during cell respiration (see pp. 22–3). The waste product of this process, carbon dioxide, is delivered to the lungs, breathed out, and then used by plants for photosynthesis.

Making sounds

FROM THE MOMENT of birth, humans make
sounds. At first these sounds are inarticulate
noises, but, as children grow, they learn to
articulate, producing distinctive words and phrases.
This ability to communicate with a voice – whether
by speaking, singing, shouting, or whispering – is
unique to humans. Sounds originate in the larynx,
or voice box, which links the base of the pharynx
(throat) to the trachea. Air expelled from the
lungs makes vocal cords in the larynx vibrate
and produce sounds, which are amplified by the
pharynx, and shaped into recognizable words
by the tongue and lips. The whole process is
controlled by several parts of the brain, notably
Broca's area, found on the left side of the brain.

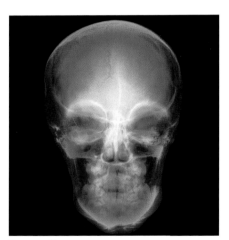

COLOURED X-RAY OF HUMAN SKULL

IMPROVING QUALITY
Sounds produced by
the larynx are improved
by other parts of the
respiratory system. In
particular, the pharynx,
nasal cavities, mouth, and
sinuses (hollow chambers
in the bones of the skull,
seen as black triangles
below the eye sockets
in this X-ray) act as
resonators. They help
to amplify the voice and
give it a distinctive quality,
in a manner similar to
the soundbox of a guitar.

VOCAL CORDS

These membrane folds extend horizontally
from the front to the back of the larynx.
Muscles attached to the vocal cords alter both
their length and the size of the opening
between them. To make sounds, the vocal
cords are drawn together, and air is expelled
from the lungs. As the air pushes between the
vocal cords, they vibrate, creating sounds.
Tightly stretched vocal cords vibrate rapidly to
produce high-pitched sounds, while looser
cords produce low-pitched sounds. Men have
lower pitched voices than women because their
vocal cords are longer and thicker and vibrate
more slowly. Loudness depends on the force
with which air passes between the cords –
the greater the force, the louder the sound.

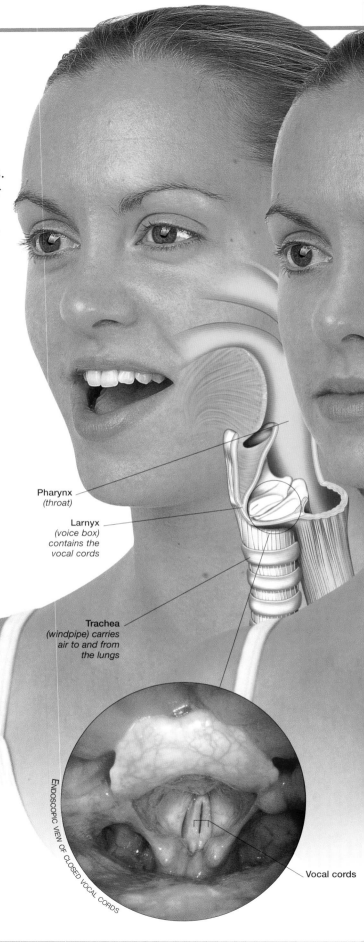

Pharynx
(throat)

Larnyx
(voice box)
contains the
vocal cords

Trachea
(windpipe) carries
air to and from
the lungs

ENDOSCOPIC VIEW OF CLOSED VOCAL CORDS

Vocal cords

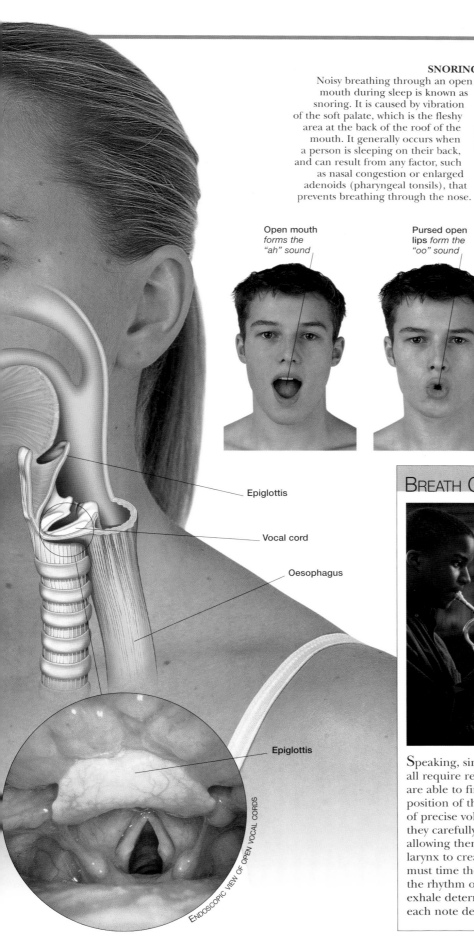

SNORING

Noisy breathing through an open mouth during sleep is known as snoring. It is caused by vibration of the soft palate, which is the fleshy area at the back of the roof of the mouth. It generally occurs when a person is sleeping on their back, and can result from any factor, such as nasal congestion or enlarged adenoids (pharyngeal tonsils), that prevents breathing through the nose.

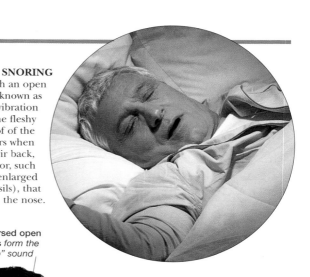

Open mouth *forms the* "ah" sound

Pursed open lips *form the* "oo" sound

SHAPING SPEECH

In a process called articulation, muscles controlling the tongue, cheeks, and lips shape the basic sounds arriving from the throat into recognizable vowels and consonants in order to form the subtle and varied sounds of speech. For example, separation of the lips so that air suddenly escapes from between them produces the sound "p".

Epiglottis

Vocal cord

Oesophagus

Epiglottis

ENDOSCOPIC VIEW OF OPEN VOCAL CORDS

BREATH CONTROL

Speaking, singing, and the playing of wind instruments all require remarkable breathing control. Singers are able to finely co-ordinate the shape, tension, and position of their vocal cords in order to produce sounds of precise volume, pitch, and quality. In addition, they carefully control the pattern of their breathing, allowing them to regulate the flow of air through the larynx to create the desired sounds. Similarly, musicians must time their breathing so that it does not interrupt the rhythm of the music, and the way in which they exhale determines whether the sound is staccato (with each note detached) or prolonged, soft or powerful.

DIGESTIVE system

DURING THEIR lifetime, the average person eats their way through at least 20 tonnes of food. The job of the digestive system is to turn this mountain of nourishment into substances that the body can use, both for energy and for growth and repair. It works like an assembly line in reverse, turning complex nutrients into simpler ones, which the body can then absorb. Food contains three major kinds of nutrient – carbohydrates, fats, and proteins – and the digestive system deals with each kind in a different way. Once these and other nutrients have been extracted, the system gets rid of any undigested waste.

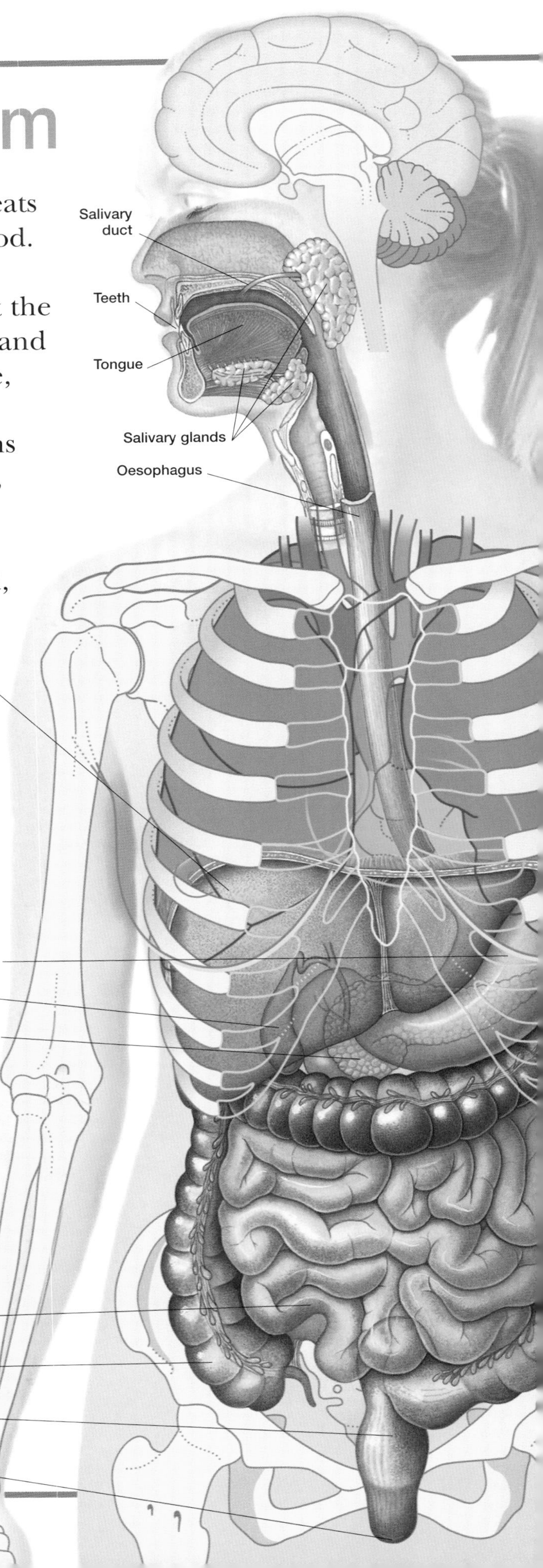

ESSENTIALS AND ACCESSORIES

The core of the digestive system is a long tube called the alimentary canal, or gastrointestinal tract. It runs from the mouth to the anus, and is divided into distinct regions – the oesophagus and stomach, and the small and large intestines – which carry out different tasks. Attached to the tube are a number of accessory organs that help in the process of digestion. They include the teeth, tongue, and salivary glands, as well as the liver, gall bladder, and pancreas. Cells lining the alimentary canal experience a lot of wear and tear, and they often have a working life of just three or four days.

CHEMICAL AND MECHANICAL
This computer-generated image shows a single molecule of pepsin – an enzyme, or chemical catalyst, that breaks proteins down into smaller units called peptides. Pepsin is produced in the stomach, and is one of more than a dozen enzymes that play a part in digestion. The mechanical side of digestion includes chewing and also muscular "churning" by the stomach and intestines. This helps enzymes to get at the substances that they break down.

Salivary duct

Teeth

Tongue

Salivary glands

Oesophagus

Liver

Stomach

Gall bladder

Pancreas

Small intestine

Large intestine

Rectum

Anus

FOOD ON THE MOVE

The alimentary canal operates by muscle power and works at a range of different speeds. Food is only briefly in the mouth and oesophagus, but once it reaches the stomach, it can stay there several hours. From here, food moves into the small intestine, where it is nudged through the lengthy twists and turns at about 1 cm (0.4 in) per minute. Once in the large intestine, it slows down again, particularly if the body is short of water. The times shown below are for a typical meal.

00:00:10

Food enters the stomach about ten seconds after it has been swallowed

03:00:00

If food contains only a small amount of fat, it leaves the stomach within three hours. Fatty or protein-rich food can stay in the stomach for twice as long as this

06:00:00

Semi-digested food, *called chyme, reaches the halfway point of the small intestine about three hours after it left the stomach. By now, many of its nutrients have been absorbed*

Small intestine

08:00:00

Ileocaecal sphincter (beginning of large intestine)

About eight hours after being swallowed, watery, indigestible waste completes its journey to the end of the small intestine

Large intestine has a 2 cm (0.8 in) layer of bacteria

20:00:00

By the time it reaches the midpoint of the large intestine, a large proportion of the waste's water has been removed and reabsorbed

During the 12 to 36 hours in the large intestine liquid waste is transformed into semi-solid faeces

32:00:00

Faeces *reach the rectum, the end of the large intestine, between 20 and 44 hours after swallowing*

DIGESTING FOOD

Although food is full of nutrients, most of them are complex molecules that the body cannot absorb. These have to be broken down into smaller and simpler chemicals, which can travel through the lining of the small intestine and into the body itself. These simple substances are formed by enzymes. Enzymes work like chemical scissors, cutting up the large molecules at specific points.

Polysaccharides (starch)

Salivary and pancreatic amylase

Disaccharides (maltose)

Maltase

Monosaccharides (glucose)

Carbohydrate digestion

Complex carbohydrates, such as starch, are broken down by enzymes in saliva and in the small intestine. An enzyme called amylase splits long starch molecules to produce maltose. Maltase then splits maltose molecules to produce glucose. Through this process, long carbohydrate molecules, or polysaccharides, are turned into monosaccharides, or simple sugars.

Protein digestion

The first step in protein digestion takes place in the stomach, where pepsin breaks down protein molecules into smaller units called peptides. In the small intestine, an enzyme called trypsin continues this work, while other enzymes, called peptidases, cut up the peptide molecules to produce individual amino acids.

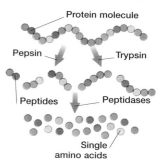

Protein molecule

Pepsin

Trypsin

Peptides

Peptidases

Single amino acids

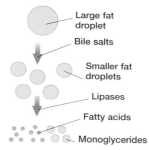

Large fat droplet

Bile salts

Smaller fat droplets

Lipases

Fatty acids

Monoglycerides

Fat digestion

Fats do not normally dissolve in water, but bile salts turn them into an emulsion of droplets. They are then digested by an enzyme called lipase, which produces fatty acids and monoglycerides. These travel to the sides of the small intestine in microscopic globules, called micelles.

DIGESTIVE SYSTEM FUNCTIONS

Ingestion	*Takes food and drink into the body through the mouth.*
Propulsion	*Moves food and indigestible waste through the alimentary canal by waves of muscle contraction (peristalsis).*
Mechanical digestion	*Physically breaks down food by chewing, and by muscular movements of the alimentary canal.*
Chemical digestion	*By using enzymes, breaks down complex nutrient molecules into simpler substances that the body can absorb.*
Absorption	*Moves digested nutrients from the alimentary canal to the bloodstream or the lymphatic system so that they can be distributed to the body's cells.*
Egestion	*During defecation, disposes of indigestible waste and waste products produced by the body.*

Teeth

LOCATED AT THE ENTRANCE of the digestive system, teeth are the hardest objects in the body. They can withstand tremendous pressure when they bite – thanks not only to their extra-tough crowns, but also to their shock-absorbing roots. They cut, crush, and chew the food that we eat, making it easier both to swallow and digest. Humans have two sets of teeth, and in each, different teeth carry out different work. But teeth all share one important characteristic: once they have appeared, or "erupted", above the gums, their hard outer enamel cannot be repaired or replaced. Enamel can be damaged by acids from food and, if it is breached, the inner part of teeth can decay. But with regular cleaning and a healthy diet, adult teeth can last for life.

TYPES OF TEETH

When someone opens their mouth wide, the differences between individual teeth become easy to see. The incisors, at the front, are the only teeth that have a flat cross-section, with a single cutting edge. They take large chunks out of food, and slice it up. The canines, to each side of them, have a single point, for gripping and tearing. Behind them are the premolars and molars, which are used for chewing food, grinding it down into a paste. These teeth have two or four cusps, or raised edges, and because they are near the back of the jaw, they have an exceptionally powerful bite.

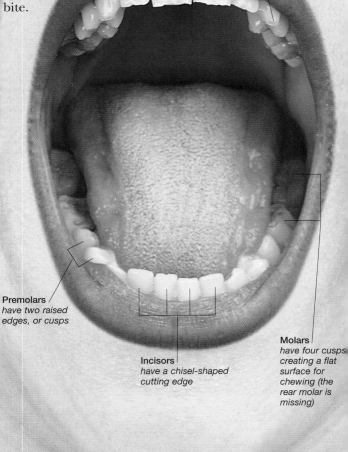

Canines *end in a single rounded point*

Premolars *have two raised edges, or cusps*

Incisors *have a chisel-shaped cutting edge*

Molars *have four cusps creating a flat surface for chewing (the rear molar is missing)*

TWO SETS

Milk teeth, also known as deciduous teeth, have small crowns and relatively shallow roots. They begin to appear at about the age of six months, and are complete by about 32 months. From the age of about six years onwards, they are shed and replaced by adult or permanent teeth, which are larger, with longer roots. Most people have 20 milk teeth and 32 adult teeth. Milk teeth appear in a set order, starting with the central incisors and ending with the second molars. Adult teeth start appearing at the front of the jaw, and work backwards as the jaw grows. However, in some people, the third molars, or "wisdom teeth", remain embedded in the jaws.

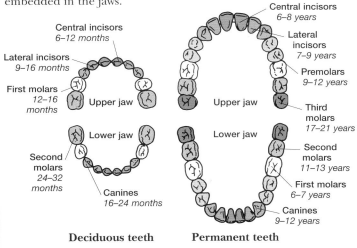

Central incisors *6–12 months*

Lateral incisors *9–16 months*

First molars *12–16 months*

Upper jaw

Lower jaw

Second molars *24–32 months*

Canines *16–24 months*

Deciduous teeth

Central incisors *6–8 years*

Lateral incisors *7–9 years*

Premolars *9–12 years*

Third molars *17–21 years*

Upper jaw

Lower jaw

Second molars *11–13 years*

First molars *6–7 years*

Canines *9–12 years*

Permanent teeth

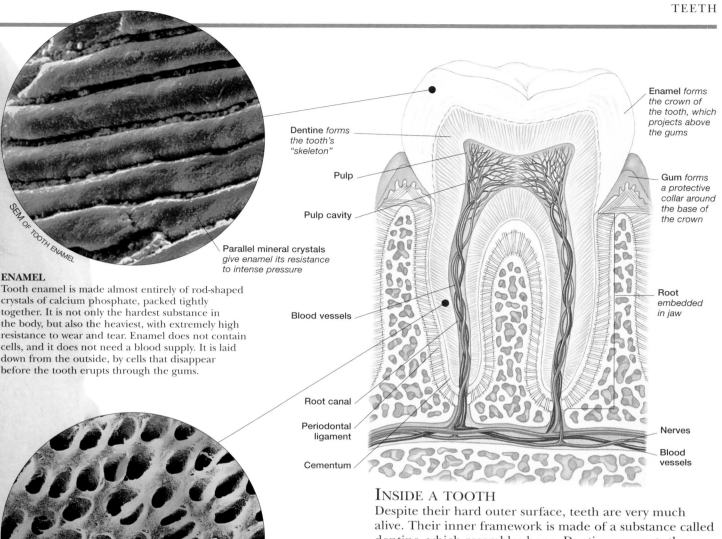

Enamel *forms the crown of the tooth, which projects above the gums*

Dentine *forms the tooth's "skeleton"*

Pulp

Pulp cavity

Gum *forms a protective collar around the base of the crown*

Blood vessels

Root *embedded in jaw*

Root canal

Periodontal ligament

Cementum

Nerves

Blood vessels

SEM OF TOOTH ENAMEL

Parallel mineral crystals *give enamel its resistance to intense pressure*

ENAMEL
Tooth enamel is made almost entirely of rod-shaped crystals of calcium phosphate, packed tightly together. It is not only the hardest substance in the body, but also the heaviest, with extremely high resistance to wear and tear. Enamel does not contain cells, and it does not need a blood supply. It is laid down from the outside, by cells that disappear before the tooth erupts through the gums.

SEM OF DENTINE

Struts *in dentine contain a higher proportion of minerals than in bone, making it harder*

INSIDE A TOOTH
Despite their hard outer surface, teeth are very much alive. Their inner framework is made of a substance called dentine, which resembles bone. Dentine supports the tooth's enamel crown, and it also forms the roots that anchor the tooth in the jaw. The roots are held in place by a chemical cement, which glues them to a ligament lining the socket. At the centre of the tooth is a natural cavity, which is full of living tissue, called pulp. This contains blood vessels and nerves, which reach the cavity through hollow root canals. The nerves enable teeth to sense changes in temperature, and, unfortunately, pain.

DENTINE
Compared to enamel, dentine has a more open structure, but it is still heavier and harder than most types of bone. When a tooth bites, it acts like crumple-resistant scaffolding, transmitting the force of the bite between the crown and the jaw. Unlike enamel, dentine is produced by cells in the pulp cavity, and it needs a blood supply to stay alive.

Bacteria *release acids when they break down the sugars in fragments of food*

PROBLEMS WITH PLAQUE
Magnified about 100 times, this photograph shows a layer of plaque on the surface of a tooth. Plaque is a mixture of bacteria and food that builds up on teeth that are not properly cleaned. Bacteria in plaque release acids as they feed, and these can eat through tooth enamel, creating holes that lead to the pulp cavity inside. If this is left untreated, the result is dental caries or tooth decay – an infection that destroys pulp cells, and dentine as well. Plaque can be hard to remove, because many bacteria produce a sticky "glue" that fastens them in place.

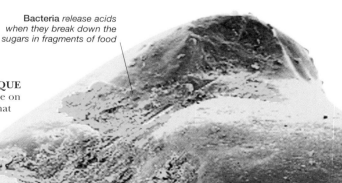

Chewing and swallowing

COMPARED WITH SOME ANIMALS, people are slow eaters. The human digestive system is not designed to cope with food in large chunks, so instead, it has to be chewed. Chewing grinds food down into a more manageable form, so that it can be swallowed. Chewing and swallowing happen almost without our noticing, but they both involve complicated movements and rapid reflexes. The tongue manoeuvres food into position between the teeth, while making sure that it does not get bitten itself. The teeth close with exactly the right amount of force, but they immediately stop if they touch something unexpectedly hard. Once the food has been reduced to a pulp, the tongue pushes it to the back of the throat. This triggers swallowing, which sends another mouthful on its way to the stomach.

LM OF A SECTION THROUGH A SALIVARY GLAND

Parotid salivary gland *located in front of the ear*

Salivary duct *carries saliva into the mouth*

Teeth *break up food*

Tongue *manoeuvres food during chewing*

ENDOSCOPIC VIEW OF OESOPHAGUS

Sublingual salivary gland *is found under the tongue*

Submandibular salivary gland *is found deep in the floor of the mouth*

Entrance to the stomach

Oesophagus *conveys food to the stomach*

INSIDE THE MOUTH

The mouth is the reception centre of the digestive system, and the place where food gets its initial processing before being passed on. As soon as it arrives, food is given a rapid check by taste buds in the tongue, to make sure that it does not contain anything that might be dangerous. At the same time, the food is bathed in saliva from the three pairs of salivary glands, which moistens it so that it is easier to swallow. Saliva contains the enzyme amylase, and this starts to digest any starch that the food contains. As chewing begins, the lips and cheeks work with the tongue to help guide the food between the teeth.

INSIDE THE OESOPHAGUS

The lining of the oesophagus is coated with mucus, which helps food to slide through on its way to the stomach. Cells lining the oesophagus have tiny folds called microplicae, and these trap mucus, keeping the inner surface slippery. Unlike the trachea, or windpipe, the oesophagus is not reinforced by cartilage because it does not need to be open at all times. When it is not in use, its upper reaches – in the throat – are usually pressed flat.

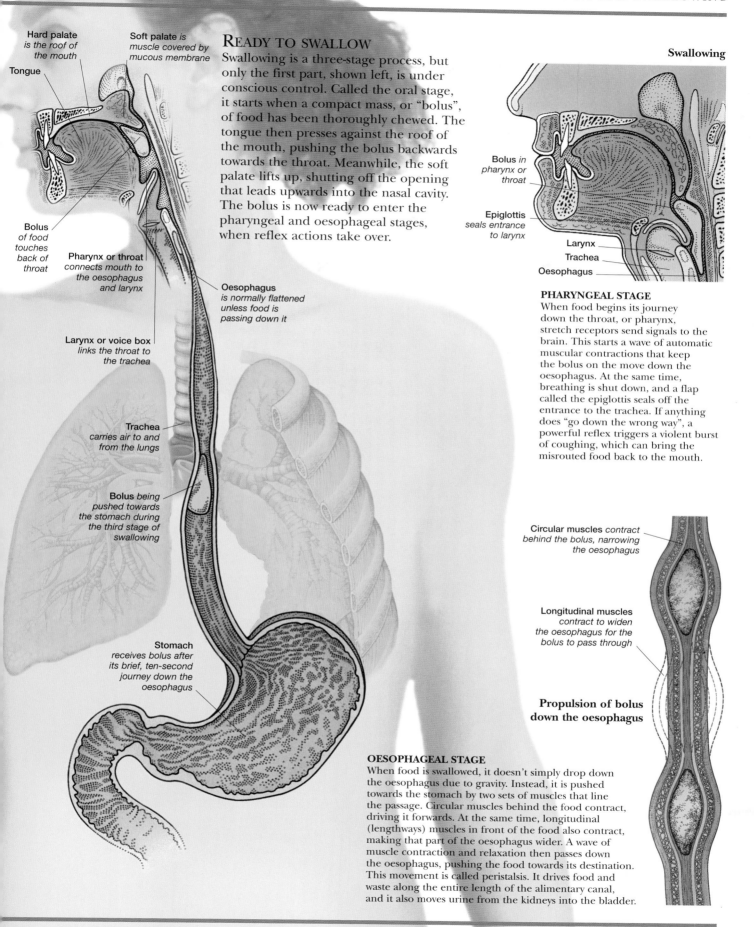

Hard palate *is the roof of the mouth*

Soft palate *is muscle covered by mucous membrane*

Tongue

Bolus *of food touches back of throat*

Pharynx or throat *connects mouth to the oesophagus and larynx*

Larynx or voice box *links the throat to the trachea*

Trachea *carries air to and from the lungs*

Bolus *being pushed towards the stomach during the third stage of swallowing*

Stomach *receives bolus after its brief, ten-second journey down the oesophagus*

Oesophagus *is normally flattened unless food is passing down it*

READY TO SWALLOW

Swallowing is a three-stage process, but only the first part, shown left, is under conscious control. Called the oral stage, it starts when a compact mass, or "bolus", of food has been thoroughly chewed. The tongue then presses against the roof of the mouth, pushing the bolus backwards towards the throat. Meanwhile, the soft palate lifts up, shutting off the opening that leads upwards into the nasal cavity. The bolus is now ready to enter the pharyngeal and oesophageal stages, when reflex actions take over.

Swallowing

Bolus *in pharynx or throat*

Epiglottis *seals entrance to larynx*

Larynx

Trachea

Oesophagus

PHARYNGEAL STAGE

When food begins its journey down the throat, or pharynx, stretch receptors send signals to the brain. This starts a wave of automatic muscular contractions that keep the bolus on the move down the oesophagus. At the same time, breathing is shut down, and a flap called the epiglottis seals off the entrance to the trachea. If anything does "go down the wrong way", a powerful reflex triggers a violent burst of coughing, which can bring the misrouted food back to the mouth.

Circular muscles *contract behind the bolus, narrowing the oesophagus*

Longitudinal muscles *contract to widen the oesophagus for the bolus to pass through*

Propulsion of bolus down the oesophagus

OESOPHAGEAL STAGE

When food is swallowed, it doesn't simply drop down the oesophagus due to gravity. Instead, it is pushed towards the stomach by two sets of muscles that line the passage. Circular muscles behind the food contract, driving it forwards. At the same time, longitudinal (lengthways) muscles in front of the food also contract, making that part of the oesophagus wider. A wave of muscle contraction and relaxation then passes down the oesophagus, pushing the food towards its destination. This movement is called peristalsis. It drives food and waste along the entire length of the alimentary canal, and it also moves urine from the kidneys into the bladder.

Stomach

Of all the body's internal organs, the stomach is probably the best known but most misunderstood. This J-shaped bag is tucked beneath the ribs, and although it can stretch to fill with food, it does not absorb any of the nutrients that food contains. Instead, the stomach has two main functions: it gets digestion underway, storing any semi-digested food, and then releases it at a slow and steady rate. In the stomach, highly acidic gastric juice allows enzymes to break down proteins, while powerful waves of muscle contraction churn the food to mix it up. After several hours of this kind of treatment, the runny result – called chyme – is ready to move on.

Oesophagus

Lower oesophageal sphincter *closes the junction between the oesophagus and stomach to keep the stomach contents in place*

Longitudinal muscle *runs the length of the stomach*

Outer covering of stomach

Circular muscle *wraps around the stomach*

Inside the Stomach

The stomach is the widest and most elastic part of the alimentary canal. When it is empty, it can be smaller than a fist, but its volume can increase by more than 20 times after a meal, because the rugae (deep folds) of its inner surface become smoother as it fills. Unlike the rest of the alimentary canal, the stomach's lining has three layers of smooth muscle, arranged at angles to each other. By contracting in turn, these muscles churn up the food. At the base of the stomach, a ring of muscle called the pyloric sphincter acts like a valve, controlling the release of semi-digested food.

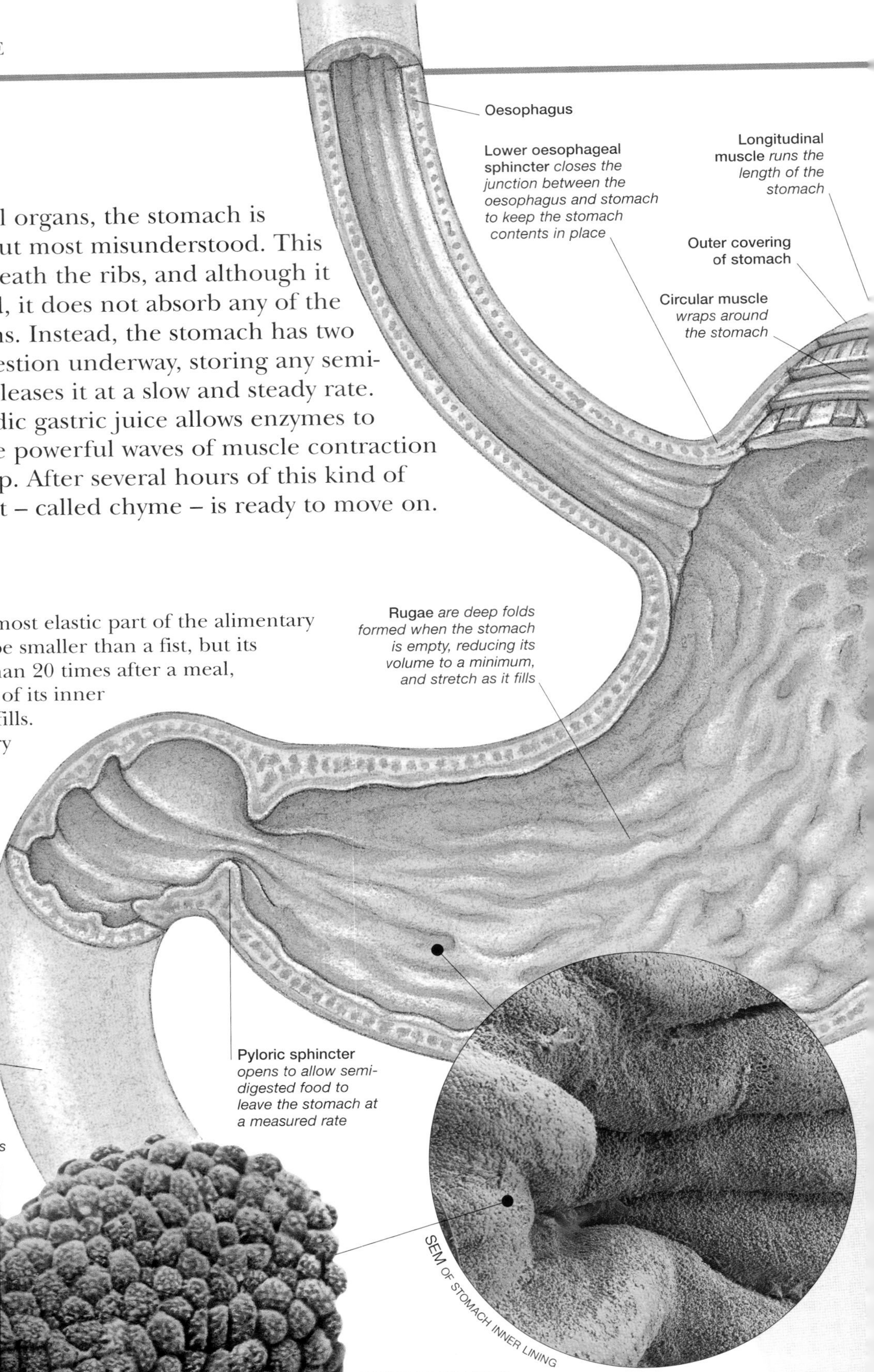

Rugae *are deep folds formed when the stomach is empty, reducing its volume to a minimum, and stretch as it fills*

Duodenum

Gastric pit *in the stomach wall leads to a gastric gland*

Pyloric sphincter *opens to allow semi-digested food to leave the stomach at a measured rate*

SEM OF STOMACH INNER LINING

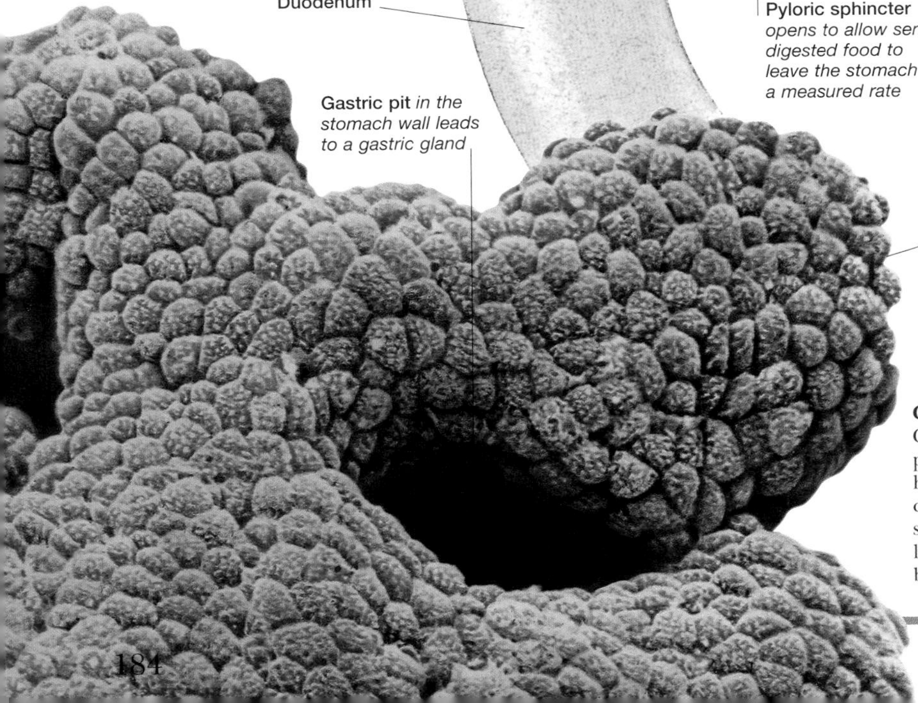

GASTRIC GLANDS

Cells in the stomach's millions of gastric glands produce the components of gastric juice – mucus, hydrochloric acid, and pepsinogen, a substance that is converted into protein-digesting pepsin as it flows into the stomach. The stomach does not digest itself because its lining is covered with protective mucus, and because pepsin becomes active only when it has been "primed" by acid.

Oblique muscle
runs diagonally

FILLING AND EMPTYING

By the time food reaches the stomach, the stomach is ready to receive it because it has been primed by the autonomic nervous system (see pp. 98–99). During its stay here – which can last for up to four hours – nerves and hormones work together to keep the digestive process working smoothly. These two control systems ensure that the stomach secretes enough gastric juice, and also trigger muscular movements (peristalsis) in the stomach wall. When digestion has progressed far enough, the pyloric sphincter relaxes, and the stomach's contents flow into the small intestine.

Pyloric sphincter *partially relaxes to allow food to pass through*

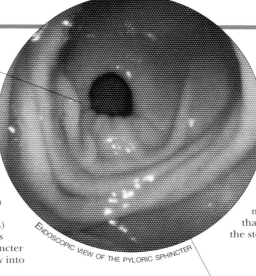

ENDOSCOPIC VIEW OF THE PYLORIC SPHINCTER

AUTOMATIC VALVE
Seen through an endoscope, the pyloric sphincter guards the entrance to the duodenum – the first part of the small intestine. When this ring of muscle is tightly contracted, nothing can leave the stomach, but as digestion proceeds, it begins to relax. This relaxation is controlled by a feedback mechanism, which ensures that semi-digested food leaves the stomach at the right rate.

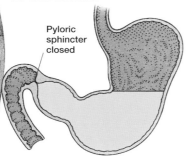

Pyloric
sphincter
closed

Filling
As it fills, the stomach releases gastric juice, which is mixed with food by waves of muscular contraction, or peristalsis.

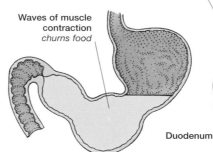

Waves of muscle
contraction
churns food

Digestion
Vigorous peristalsis churns food as gastric juice digests it into creamy liquid chyme.

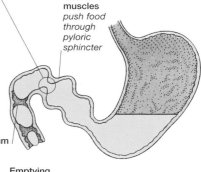

Stomach
muscles
*push food
through
pyloric
sphincter*

Duodenum

Emptying
If chyme is liquid enough, the pyloric sphincter relaxes and opens slightly to let small quantities of food pass into the duodenum.

Stomach wall
*contains millions
of microscopic glands
that secrete gastric juice*

ACID ATTACK

When food enters the stomach, it is mixed with a digestive fluid called gastric juice. This juice contains hydrochloric acid, and it is strong enough to dissolve small pieces of bone. These acidic conditions are needed for protein digestion, which is carried out by an enzyme called pepsin.

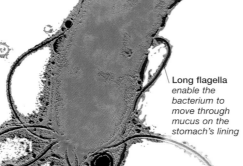

Long flagella
*enable the
bacterium to
move through
mucus on the
stomach's lining*

STOMACH BUG
Stomach acid kills most bacteria, but one kind, called *Helicobacter pylori*, manages to thrive in these hazardous conditions. In recent years, this bacterium has been closely studied, and there is mounting evidence that it is linked to two different forms of stomach disease, one of which is stomach cancer. The other disease is gastritis – an inflammation that often leads to ulcers. How the bacterium spreads is not known.

STOMACH STUDY

In 1822, American surgeon William Beaumont treated Alexis St Martin, who had been shot during a hunting trip. Beaumont saved the man's life, but his patient was left with a permanent opening from his stomach to the outside. For the next decade, Beaumont monitored St Martin's stomach, and the fluid that it produced. Although gruesome, the research produced a great deal of useful information, and the patient lived to the ripe age of 82.

Beaumont (1785–1853) examining St Martin's stomach

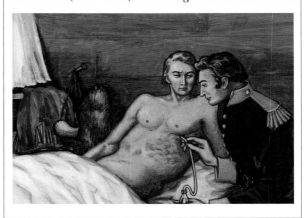

Small intestine

DESPITE ITS MODEST-SOUNDING NAME, the small intestine is the most important part of the entire digestive system. Measuring up to 6 m (19.7 ft) in length, this intricately folded tube is the site where food is fully broken down, and where its nutrients are absorbed. By the time food has been gently squeezed through all its twists and turns, nearly all its useful ingredients have been removed, leaving just watery waste. The small intestine is only about 2.5 cm (1 in) across, but its inner lining has a huge surface area, thanks to microscopic projections called villi. With the help of the pancreas and the liver, the intestine breaks down food into simple substances, and the villi absorb these into the body itself.

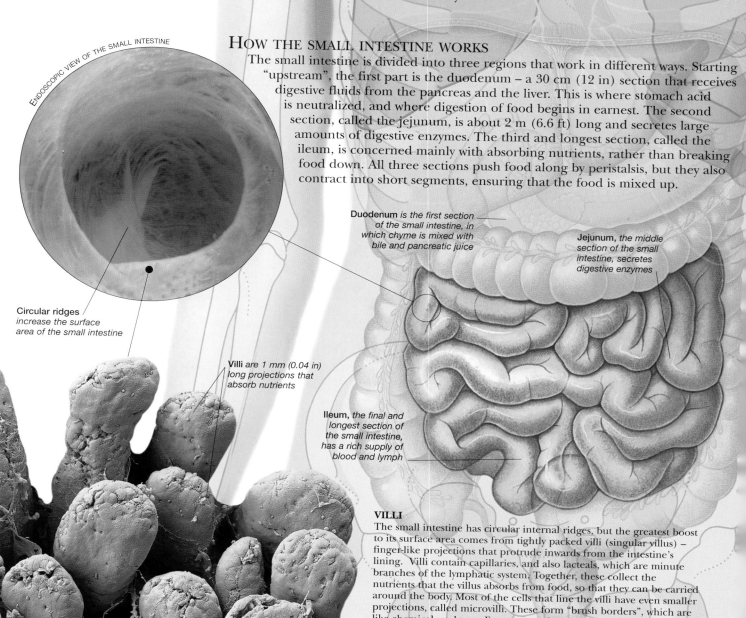

ENDOSCOPIC VIEW OF THE SMALL INTESTINE

HOW THE SMALL INTESTINE WORKS

The small intestine is divided into three regions that work in different ways. Starting "upstream", the first part is the duodenum – a 30 cm (12 in) section that receives digestive fluids from the pancreas and the liver. This is where stomach acid is neutralized, and where digestion of food begins in earnest. The second section, called the jejunum, is about 2 m (6.6 ft) long and secretes large amounts of digestive enzymes. The third and longest section, called the ileum, is concerned mainly with absorbing nutrients, rather than breaking food down. All three sections push food along by peristalsis, but they also contract into short segments, ensuring that the food is mixed up.

Duodenum *is the first section of the small intestine, in which chyme is mixed with bile and pancreatic juice*

Jejunum, *the middle section of the small intestine, secretes digestive enzymes*

Circular ridges *increase the surface area of the small intestine*

Villi *are 1 mm (0.04 in) long projections that absorb nutrients*

Ileum, *the final and longest section of the small intestine, has a rich supply of blood and lymph*

VILLI

The small intestine has circular internal ridges, but the greatest boost to its surface area comes from tightly packed villi (singular villus) – finger-like projections that protrude inwards from the intestine's lining. Villi contain capillaries, and also lacteals, which are minute branches of the lymphatic system. Together, these collect the nutrients that the villus absorbs from food, so that they can be carried around the body. Most of the cells that line the villi have even smaller projections, called microvilli. These form "brush borders", which are like chemical worktops. Enzymes are fastened to the brush borders, and they carry out the last stages of digestion before food is absorbed.

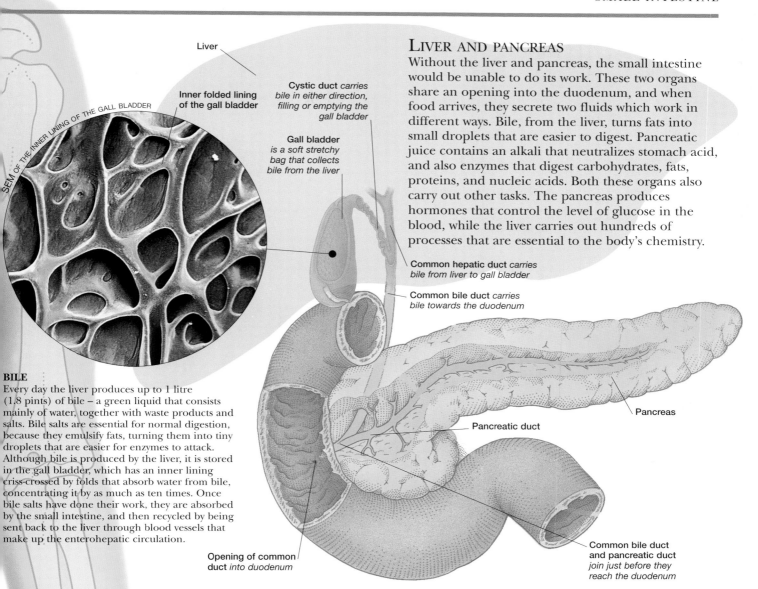

Liver

SEM OF THE INNER LINING OF THE GALL BLADDER

Inner folded lining
of the gall bladder

Cystic duct *carries
bile in either direction,
filling or emptying the
gall bladder*

Gall bladder
*is a soft stretchy
bag that collects
bile from the liver*

Common hepatic duct *carries
bile from liver to gall bladder*

Common bile duct *carries
bile towards the duodenum*

Pancreas

Pancreatic duct

Common bile duct
and pancreatic duct
*join just before they
reach the duodenum*

Opening of common
duct *into duodenum*

LIVER AND PANCREAS

Without the liver and pancreas, the small intestine would be unable to do its work. These two organs share an opening into the duodenum, and when food arrives, they secrete two fluids which work in different ways. Bile, from the liver, turns fats into small droplets that are easier to digest. Pancreatic juice contains an alkali that neutralizes stomach acid, and also enzymes that digest carbohydrates, fats, proteins, and nucleic acids. Both these organs also carry out other tasks. The pancreas produces hormones that control the level of glucose in the blood, while the liver carries out hundreds of processes that are essential to the body's chemistry.

BILE

Every day the liver produces up to 1 litre (1.8 pints) of bile – a green liquid that consists mainly of water, together with waste products and salts. Bile salts are essential for normal digestion, because they emulsify fats, turning them into tiny droplets that are easier for enzymes to attack. Although bile is produced by the liver, it is stored in the gall bladder, which has an inner lining criss-crossed by folds that absorb water from bile, concentrating it by as much as ten times. Once bile salts have done their work, they are absorbed by the small intestine, and then recycled by being sent back to the liver through blood vessels that make up the enterohepatic circulation.

ENZYMES IN ACTION

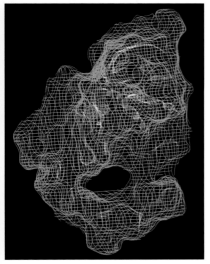

**Computer-generated image
of the enzyme amylase**

Enzymes are proteins that speed up chemical reactions in the body. If they did not exist, it would take several decades to digest even a single meal. There are two main groups of digestive enzymes: those made by the pancreas and released in pancreatic juice, and those on the "brush borders" of the small intestine. Each enzyme acts on one particular type of nutrient, turning it into smaller and simpler products. Enzyme molecules are not used up when they do their work. This means that they can do the same job thousands or even millions of times in succession.

ENZYME	ACTS ON	PRODUCT
PANCREATIC		
Amylase	Starch	Maltose
Trypsin	Proteins	Peptides
Chymotrypsin	Proteins	Peptides
Carboxypeptidase	Proteins	
Lipase	Fats and oils	Fatty acids and monoglycerides
Nuclease	Nucleic acids	Pentoses and bases
BRUSH BORDER		
Peptidases	Peptides	Amino acids
Maltase	Maltose	Glucose
Sucrase	Sucrose	Glucose and fructose
Lactase	Lactose	Glucose and galactose
Nuclease	Nucleic acids	Pentoses and bases

Large intestine

THE LARGE INTESTINE IS the final stretch of the alimentary canal. It is more than twice as wide as the small intestine, but only about one-quarter as long. Instead of twisting and turning, it follows a more straightforward path, with just a handful of sharp bends. The large intestine does not produce any enzymes, and it does not play a direct part in digestion. Instead, its chief function is to reabsorb water to help the body's fluid balance, and to make waste easier to expel. The large intestine also has another role: it absorbs vitamins that are made by bacteria. Huge numbers of microbes thrive in its warm and moist interior, and they break down substances that have escaped digestion, before eventually being expelled themselves.

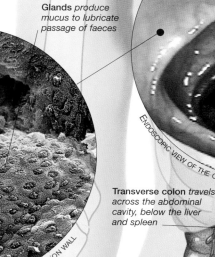
SEM OF COLON WALL

Glands *produce mucus to lubricate passage of faeces*

INSIDE THE COLON
Endoscopes are often used to examine the colon for signs of disease. This view shows the inside of a healthy colon, with a corridor of pockets, called haustra, separated by narrower parts of the intestine wall. The intestine looks triangular in cross-section because it has three taeniae, (bands of muscle) running almost all the way along it.

ENDOSCOPIC VIEW OF THE COLON

Descending colon *travels down the left side of the abdominal cavity*

Taenia omentalis *is one of three parallel bands of muscle that run along the colon*

Transverse colon *travels across the abdominal cavity, below the liver and spleen*

ABSORPTIVE LINING
Magnified about 400 times, the lining of the large intestine (above) looks much smoother than other parts of the alimentary canal. Unlike the small intestine, it does not have villi, but it does have small glands, which can clearly be seen in the picture (blue). These contain cells that produce mucus. Water-absorbing cells are spread all over the large intestine's lining, and in the sides of its glands.

Ascending colon *travels up the right side of the adominal cavity*

COLON
The large intestine begins at the ileocaecal sphincter, or valve, and it ends at the rectum and anus. The section between these points is called the colon and measures about 1.5 m (5 ft) in length. The colon follows a path shaped like the edge of a shield, travelling up, across, and then down the lower part of the abdominal cavity. Unlike the small intestine, the colon has muscle bands (taeniae), which gather it up into a series of pockets (haustra) that help to compact waste before passing it on. The rectum collects waste once most of its water has been removed, and holds the waste ready for disposal.

Appendix

Small intestine

Anus

Sigmoid colon *leads down to the rectum*

Rectum

MALFORMED BONES
A girl with rickets, caused by a lack of vitamin D, pictured with her brother in Budapest, Hungary, around 1920. She shows the characteristic bowing of the legs caused by her upper body weight pushing downwards on weakened leg bones.

contained an "essential food factor". Eijkmann's work was paralleled by that started in 1906 by Frederick Gowland Hopkins (1861–1947). By feeding rats carefully controlled diets, Hopkins showed that to stay healthy they – and presumably humans too – needed tiny amounts of what he termed "accessory food factors".

NAMING VITAMINS

"Food factors" were given a new name – "vitamins" – in 1912 by Polish-American chemist Casimir Funk (1884–1967). Funk's idea that diseases such as beriberi and scurvy are caused by the lack of a particular vitamin set in motion a new era of

scientific research. In 1914, Joseph Goldberger (1874–1929) of the US Public Health Service demonstrated that pellagra – a disease that causes dermatitis, diarrhoea, and dementia – was not spread by insects but was the result of poor diet, and could be reversed by a vitamin (niacin) found in protein-rich foods. British physician Edward Mellanby (1884–1955) showed in 1918 that a substance in cod liver oil – later identified by American scientist E.V. McCollum (1879–1967) as vitamin D – could prevent rickets, a deficiency disease characterized by weakened bones.

VITAMIN SUPPLEMENTS

By the end of the 1930s, scientists had identified vitamins A, C (the scurvy-preventing vitamin in citrus fruits), D, E, and the B vitamins, including B_1, B_2, and B_{12}. This wealth of knowledge about vitamins means that deficiency diseases have been all but eliminated in the developed world.

EXTRA VITAMINS
During the Second World War, food was rationed in Britain, but the government took measures to protect young children from vitamin deficiencies. Here, mothers in wartime London receive free vitamin supplements in the form of orange juice (vitamin C) and cod liver oil (vitamins A and D) for their children.

SUPPLEMENTS
A wide range of vitamin supplements like these are available from pharmacies. How effective vitamin supplements are is the subject of considerable debate. Many nutritionists believe that, for most people, a mixed diet containing fresh produce provides all the vitamins they need.

Vitamins and minerals

ESSENTIAL TO ANY BALANCED DIET are the micronutrients – minerals and vitamins – needed for growth, vitality, and general well-being. All minerals, and most vitamins, can only be obtained from food, and are not made by the body. Minerals, such as calcium, are chemical elements. Seven are required in relatively large amounts, while others – trace minerals – including iron, are needed only in minute amounts. Vitamins are organic (carbon-containing) substances. Without them the body cannot utilize macronutrients – carbohydrates, proteins, and fats – to supply energy or for building and growth. Vitamins are classified according to whether they dissolve in fats (A, D, E, K) or in water (B complex, C). The lists below reveal their major sources, functions, and the symptoms of a deficiency (shortage).

MICRONUTRIENTS
A diet that contains a wide variety of foods, including plenty of fresh fruit and vegetables, will provide all the micronutrients needed by humans. Avocado, for example, is a fruit that is rich in the mineral potassium and the vitamin folic acid. It also contains the minerals calcium, iron, magnesium, phosphorus, sodium, zinc, copper, and manganese, as well as the vitamins A, B_1, B_2, B_6, niacin, pantothenic acid, and vitamin C.

WATER-SOLUBLE VITAMINS

VITAMIN	SOURCE	ROLE IN BODY	DEFICIENCY SYMPTOMS
B_1 (THIAMINE)	Whole grains, peas, beans, nuts, yeast, egg white, fish, liver, milk	Needed by enzymes that break down sugars, and for normal nerve and muscle function	Beriberi (disease causing nerve inflammation and muscle weakness)
B_2 (RIBOFLAVIN)	Milk, cheese, liver, meat, egg white, leafy green vegetables, whole grains, peas, and beans	Needed by enzymes involved in carbohydrate and protein metabolism	Cracked skin, defective vision, cataracts, ulceration of the cornea
NIACIN (B_3, NICOTINAMIDE)	Meat, fish, wholegrain cereals, liver, peanuts, yeast, leafy green vegetables, potatoes	Plays part in breakdown of carbohydrates and fats during cell respiration	Pellagra – a disease causing skin disorders, diarrhoea, and dementia
B_6 (PYRIDOXINE)	Meat, poultry, fish, liver, wholegrain cereals, egg yolks, potatoes, spinach	Needed by enzymes involved in amino acid and fatty acid metabolism	In children, anaemia, convulsions; in adults, sores around the eyes and nose
B_{12} (CYANOCOBALAMIN)	Animal foods – meat, liver, kidney, poultry, fish, milk, eggs, oysters – and yeast	Needed by enzymes involved in synthesis of DNA and proteins; promotes red blood cell formation	Pernicious anaemia – a disease causing paleness, weight loss, and impaired nervous system function
PANTOTHENIC ACID	Meat, liver, egg yolks, wholegrain cereals, peas, beans, yeast	Needed by enzymes involved in cell respiration, fatty acid metabolism; also involved in synthesis of steroid hormones	Disorders of the digestive and nervous systems
BIOTIN	Liver, egg yolks, wholegrain cereals, yeast, peas, beans, nuts	Involved in carbohydrate, fat, and amino acid metabolism	Skin disorders, muscle pain, tiredness, depression, nausea
FOLIC ACID (FOLATE, FOLACIN)	Leafy green vegetables, liver, whole grains, eggs, yeast	Needed by enzymes involved in amino acid and DNA synthesis; involved in red blood cell formation	Digestive system disorders, anaemia
C (ASCORBIC ACID)	Citrus fruits, green peppers, tomatoes, other fruits, broccoli, fresh potatoes	Promotes collagen formation and growth of teeth, bones, and blood vessels; aids normal wound healing	Poor bone growth and wound healing; in severe cases, scurvy (bleeding gums, anaemia, weight loss, internal bleeding)

FAT-SOLUBLE VITAMINS

VITAMIN	SOURCE	ROLE IN BODY	DEFICIENCY SYMPTOMS
A (RETINOL)	Dark green leafy vegetables, yellow-orange vegetables and fruit, egg yolks, oily fish, milk, liver	Needed for production of pigments in the eye, bones, and teeth, and for healthy skin and linings of organs	Night blindness, clouding of the cornea, dry skin and hair, lowered resistance to infection
D (CHOLECALCIFEROL)	Primary source is action of sunlight on the skin; also oily fish, egg yolks, dairy products	Aids absorption of calcium from intestines; helps bone and teeth formation	Rickets (disease where bones do not develop properly)
E (TOCOPHEROLS)	Vegetable oils, leafy green vegetables, wholegrain cereals, egg yolks, liver	Formation of red blood cells; protects cell membranes from damage	Damaged red blood cells
K (PHYLLOQUINONE)	Made by bacteria in large intestine; also pork liver, cauliflower, cabbage	Needed to make substances involved in blood clotting	Easy bruising and bleeding

KEY MINERALS

MINERAL	SOURCE	ROLE IN BODY	DEFICIENCY SYMPTOMS
CALCIUM (Ca)	Milk, dairy products, shellfish, fish, leafy green vegetables, egg yolks, nuts, and seeds	Helps build healthy bones and teeth; involved in muscle contraction, nerve impulse conduction, and blood clotting	Stunted growth and rickets (see vitamin D) in children; osteoporosis in adults
CHLORINE (Cl)	Many foods, table salt	Helps maintain balance of water and ions in blood and tissue fluid; needed to form acid in stomach juices	Muscle cramps, mental apathy
MAGNESIUM (Mg)	Wholegrain cereals, leafy green vegetables, milk, dairy products, meat, nuts	Helps build bones; involved in muscle contraction, nerve impulse conduction; needed for activity of many enzymes	Stunted growth, behavioural problems, tremors
PHOSPHORUS (P)	Eggs, fish, meat, milk and dairy products, nuts, peas, beans, wholegrain cereals	Component of bones and teeth; key part of ATP (energy storage and transfer) and DNA (genetic material)	Weak and poorly formed bones
POTASSIUM (K)	Dried fruits, bananas, nuts, beans, wholegrain cereals, green leafy vegetables, meat, milk, dairy products, fish	Helps maintain balance of water and ions in blood and tissue fluid; involved in muscle contraction, nerve impulse conduction, and regular heart rhythm	Muscle weakness, paralysis, heart failure
SODIUM (Na)	Most foods (except fruits); table salt	Helps maintain balance of water and ions in blood and tissue fluid; involved in muscle contraction and nerve impulse conduction	Nausea, muscle cramps, convulsions, mental apathy
SULPHUR (S)	Foods rich in protein: meat, eggs, milk, nuts, seeds	Essential part of many proteins	Impaired protein synthesis
TRACE MINERALS			
COPPER (Cu)	Liver, meat, shellfish, mushrooms, beans, peas, wholegrain cereals	Needed for production of the haemoglobin in red blood cells and melanin in skin	Anaemia
FLUORINE (F)	Fish, shellfish, fluoridated tap water	Needed for strong teeth and bones	Tooth decay (caries)
IODINE (I)	Fish, shellfish, iodized table salt	Needed to make thyroid hormones	Goitre, reduced metabolic rate
IRON (Fe)	Liver, red meat, shellfish, nuts, egg yolks, leafy green vegetables, wholegrain cereals	Essential for making the haemoglobin in red blood cells	Anaemia
MANGANESE (Mn)	Vegetables, fruit, nuts, wholegrain cereals	Assists action of many enzymes	Poor growth
SELENIUM (Se)	Fish, shellfish, meat, wholegrain cereals, dairy products	Antioxidant (prevents damage to cells and tissues by oxidation)	Not known
ZINC (Zn)	Fish, shellfish, meat, wholegrain cereals, beans, nuts, eggs	Constituent of several enzymes; necessary for normal growth, wound healing, sperm production	Retarded growth, learning impairment, loss of taste and smell

Metabolism

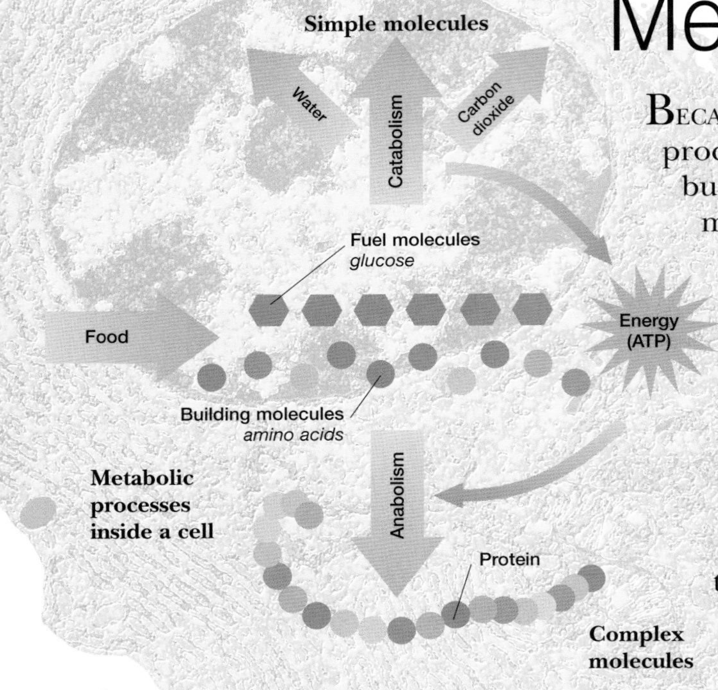

BECAUSE THE BODY IS MADE OF CELLS, it depends on chemical processes to stay alive. Exactly how many is still uncertain, but the known total already runs into many thousands, and more are being discovered all the time. Together, they make up the body's metabolism – the sum of all the chemical reactions that it carries out. Metabolism has two sides, which are closely interlinked. On the one hand, catabolic processes break down substances and release energy from them. On the other, anabolic ones take in energy, and use it to build substances and make the body work. During this energy interchange, some energy escapes, chiefly in the form of heat. The body's metabolic rate shows how fast it produces heat, and therefore how quickly it is "burning" its fuel.

Simple molecules

Water

Catabolism

Carbon dioxide

Fuel molecules
glucose

Food

Energy (ATP)

Building molecules
amino acids

Metabolic processes inside a cell

Anabolism

Protein

Complex molecules

CATABOLISM AND ANABOLISM
The diagram above shows the effect of metabolism inside a cell. In catabolism, shown in the upper half of the cell, energy-rich substances from food are oxidized in a process called cellular respiration. This releases energy that is used to drive anabolic processes (shown in the lower half of the cell). Anabolic processes make complex molecules – such as proteins – from simpler raw materials. In cells, energy is transferred between catabolic and anabolic reactions by a substance called adenosine triphosphate (ATP). This acts like a shuttle service, picking up energy when it is released, and delivering it where it is needed.

FOOD AND METABOLISM
During digestion, the complex carbohydrates, fats, and proteins in food are broken down into, respectively, glucose, fatty acids and glycerol, and amino acids. These simple nutrients are the raw materials of metabolism. This diagram reveals what happens to these nutrients after they have been processed by the liver (see pp. 198–99) and are then carried by the bloodstream to be used by body cells or stored until required.

Nutrient storage
Excess glucose is stored as glycogen in liver and muscle cells, and converted back to glucose if blood glucose levels drop. Excess fatty acids are stored as fat in adipose (fat) cells, as is glucose if glycogen stores are full. Excess amino acids cannot be stored, but can be converted to fat.

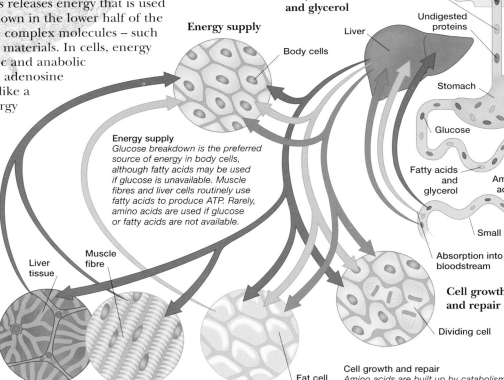

■ **Amino acids**
■ **Glucose**
■ **Fatty acids and glycerol**

Energy supply

Body cells

Liver

Undigested carbohydrates

Undigested fats

Undigested proteins

Stomach

Glucose

Energy supply
Glucose breakdown is the preferred source of energy in body cells, although fatty acids may be used if glucose is unavailable. Muscle fibres and liver cells routinely use fatty acids to produce ATP. Rarely, amino acids are used if glucose or fatty acids are not available.

Fatty acids and glycerol

Amino acids

Small intestine

Absorption into bloodstream

Cell growth and repair

Muscle fibre

Liver tissue

Dividing cell

Fat cell

Glycogen storage

Fat storage

Cell growth and repair
Amino acids are built up by catabolism into proteins, which are used in cell building, division, and repair, and for making enzymes. Fatty acids are used to make cell membranes and the myelin sheaths around nerve axons, and, with glucose, provide the energy for cell growth and repair.

SANTORIO'S BALANCE

The Italian physician Santorio (or Sanctorius) (1561–1636) devised a simple but ingenious way of investigating metabolism. He devised a balance (right) that he could sit in, and used it to measure changes in his weight after eating, sleeping, and exercise. Santorio found that the body loses weight when it is at rest, through something he called "insensible perspiration". This weight loss is a sign of catabolism – the breakdown of substances to release energy.

TAKING A TEMPERATURE

If the body is fighting an infection, its temperature often rises by 1–2°C (2–3.5°F). As a result, taking a temperature is a useful way of checking someone's health. Temperatures used to be taken with mercury thermometers, which took a minute or more to produce a reading. Now, digital thermometers give an almost instant result. A slight rise in temperature can be beneficial, because it helps to check the growth of bacteria and viruses.

Digital temperature reading *is taken in the ear*

TEMPERATURE CONTROL

Whatever the conditions outside, the body's internal temperature remains steady, at almost exactly 37°C (98.6°F). To stay at this temperature, the body has to produce heat at the same rate that it loses it. This balancing act is controlled by the hypothalamus. If the blood's temperature falls, the hypothalamus triggers actions that produce more heat, and ones that make it harder for heat to escape. If the blood's temperature rises, it cuts down heat production, and makes it easier for the body to lose some of its warmth.

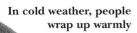

In cold weather, people wrap up warmly

The hypothalamus initiates heat loss

Heat loss
- *Blood vessels in the skin dilate (widen), increasing heat loss to the outside from blood flowing through them, and making the skin flushed*
- *Sweat glands release more sweat on to the skin's surface from where it evaporates to cool the skin*

Body temperature increases during exercise or when in a hot climate, etc

Body temperature decreases and heat loss mechanism shuts off

Normal body temperature 37°C (98.6°F)

Body temperature increases and heat gain mechanism shuts off

Body temperature decreases when in a cold climate, etc

The hypothalamus initiates heat gain

Heat gain
- *Blood vessels in the skin constrict (narrow), diverting blood away from the skin so less heat is lost to the outside*
- *Shivering (involuntary contraction of skeletal muscles) generates extra heat*
- *Erection of hairs slightly reduces heat loss from the head*

Dilated blood vessels | Hair | Sweat droplet

Sweat gland

Sections through skin

Constricted blood vessels | Erect hair

Contracted arrector pili muscle

HYPOTHERMIA AND HEATSTROKE

When it is cold, people keep warm by wearing several layers and having hot drinks (as in the picture, left); when it is hot, they tend to discard their clothes. In extremely hot or cold conditions, the body's temperature-control mechanisms can sometimes break down. If someone suffers heatstroke, their temperature can reach 41°C (106°F) or more. In the opposite condition – called hypothermia – it drops below 35°C (95°F). Both conditions can be dangerous. However, hypothermia is sometimes used during surgery, particularly in heart operations. By slowing the body's metabolism, hypothermia reduces its need for oxygen while an operation is underway.

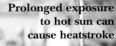

Prolonged exposure to hot sun can cause heatstroke

Liver

Wᴇɪɢʜɪɴɢ ᴀʙᴏᴜᴛ 1.4 kg (3 lb) in adults, the liver is the largest internal organ in the body. One of its jobs is to make bile (used in digestion), but its real work starts once nutrients from food have been absorbed into the blood. Its task is to process these nutrients, so that substances in the blood stay at the correct levels, keeping the body in a stable state. This mammoth operation involves more than 500 different chemical processes, which are carried out by cells called hepatocytes. Packed into columns called lobules, they are bathed by blood transported from the digestive system by the portal vein. The lobules absorb and release substances as the blood flows past. The liver also contains phagocytes (see pp. 134–5), white blood cells that remove bacteria, together with blood cells whose useful life has ended.

Pʀᴏᴄᴇssɪɴɢ ᴘʟᴀɴᴛ
The liver fits into the top of the abdominal cavity. It is lopsided, with a large right lobe and a smaller left lobe, and it partly covers the stomach. The two lobes are divided by a thick ligament, and the entire liver is covered by a sheet of tough connective tissue. The liver is unusual because it receives blood from two separate sources – the heart and the alimentary canal. It also has a remarkable ability to regenerate after injuries, growing back even if two-thirds of its cells have been lost.

Liver

Hepatic veins *empty blood into the inferior vena cava, which carries it to the heart*

Inferior vena cava

Central vein *empties blood into the hepatic vein*

Lobule

Branch of the hepatic artery

Bile duct *collects bile made by the hepatocytes*

Branch of the portal vein

Cross-section of liver lobules

Right lobe *of the liver*

Gall bladder *stores bile produced by the liver*

Common hepatic duct *carries bile to the gall bladder*

Cystic duct *carries bile in either direction, filling or emptying gall bladder*

Common bile duct *carries bile to the duodenum*

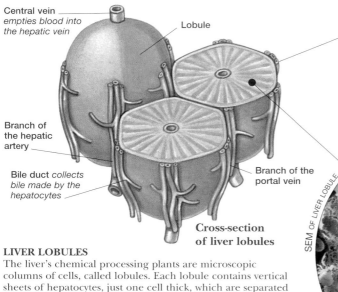

SEM OF LIVER LOBULE

Red blood cells *in a sinusoid*

LIVER LOBULES
The liver's chemical processing plants are microscopic columns of cells, called lobules. Each lobule contains vertical sheets of hepatocytes, just one cell thick, which are separated by spaces called sinusoids. Blood pumps through the sinusoids from the outside of each lobule. The hepatocytes process the blood; they remove some substances for storage, break down others, and release those that the body needs. In the sinusoids, phagocytes called Kupffer cells destroy old blood cells and bacteria. Processed blood leaves the lobule via a central vein.

HEPATOCYTES
This SEM of the inside of a sesame seed-sized lobule shows the hard-working hepatocytes, or liver cells (brown), that perform the liver's many functions. The hepatocytes surround sinusoids (blue), the leaky capillaries that provide a delivery and removal service. Between the sheets of hepatocytes are tiny canals called bile canaliculi (yellow). These carry bile secreted by hepatocytes into the bile ducts.

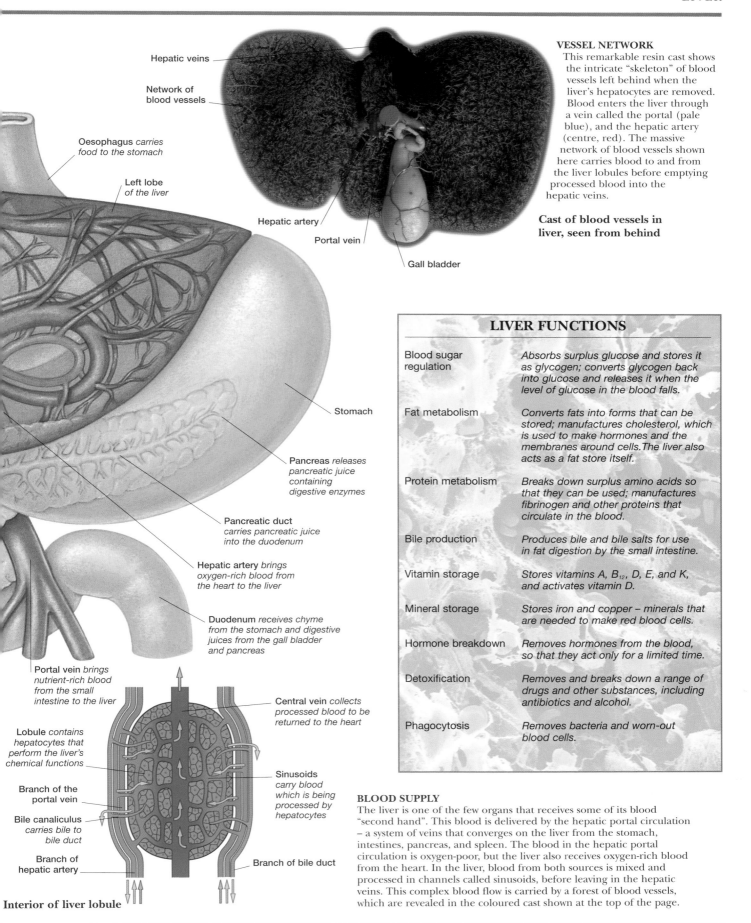

Hepatic veins

Network of
blood vessels

Oesophagus *carries
food to the stomach*

Left lobe
of the liver

Hepatic artery

Portal vein

Gall bladder

Stomach

Pancreas *releases
pancreatic juice
containing
digestive enzymes*

Pancreatic duct
*carries pancreatic juice
into the duodenum*

Hepatic artery *brings
oxygen-rich blood from
the heart to the liver*

Duodenum *receives chyme
from the stomach and digestive
juices from the gall bladder
and pancreas*

Portal vein *brings
nutrient-rich blood
from the small
intestine to the liver*

Lobule *contains
hepatocytes that
perform the liver's
chemical functions*

Branch of the
portal vein

Bile canaliculus
*carries bile to
bile duct*

Branch of
hepatic artery

Central vein *collects
processed blood to be
returned to the heart*

Sinusoids
*carry blood
which is being
processed by
hepatocytes*

Branch of bile duct

Interior of liver lobule

VESSEL NETWORK

This remarkable resin cast shows
the intricate "skeleton" of blood
vessels left behind when the
liver's hepatocytes are removed.
Blood enters the liver through
a vein called the portal (pale
blue), and the hepatic artery
(centre, red). The massive
network of blood vessels shown
here carries blood to and from
the liver lobules before emptying
processed blood into the
hepatic veins.

**Cast of blood vessels in
liver, seen from behind**

LIVER FUNCTIONS

Blood sugar regulation	*Absorbs surplus glucose and stores it as glycogen; converts glycogen back into glucose and releases it when the level of glucose in the blood falls.*
Fat metabolism	*Converts fats into forms that can be stored; manufactures cholesterol, which is used to make hormones and the membranes around cells. The liver also acts as a fat store itself.*
Protein metabolism	*Breaks down surplus amino acids so that they can be used; manufactures fibrinogen and other proteins that circulate in the blood.*
Bile production	*Produces bile and bile salts for use in fat digestion by the small intestine.*
Vitamin storage	*Stores vitamins A, B_{12}, D, E, and K, and activates vitamin D.*
Mineral storage	*Stores iron and copper – minerals that are needed to make red blood cells.*
Hormone breakdown	*Removes hormones from the blood, so that they act only for a limited time.*
Detoxification	*Removes and breaks down a range of drugs and other substances, including antibiotics and alcohol.*
Phagocytosis	*Removes bacteria and worn-out blood cells.*

BLOOD SUPPLY

The liver is one of the few organs that receives some of its blood
"second hand". This blood is delivered by the hepatic portal circulation
– a system of veins that converges on the liver from the stomach,
intestines, pancreas, and spleen. The blood in the hepatic portal
circulation is oxygen-poor, but the liver also receives oxygen-rich blood
from the heart. In the liver, blood from both sources is mixed and
processed in channels called sinusoids, before leaving in the hepatic
veins. This complex blood flow is carried by a forest of blood vessels,
which are revealed in the coloured cast shown at the top of the page.

Energy balance

Humans share one feature with machines: we need energy to make our bodies work. Energy cannot be touched or even seen, but without it, nerves cannot transmit signals, muscles cannot contract, and cells cannot divide and grow. We use energy when we are completely at rest, but our energy expenditure soars by 10 times or more when we are really on the move. The body obtains all its energy from food, and it manages its resources as carefully as someone running a bank account. Its "cash" consists of glucose in the blood – a supply of chemical energy that is instantly available to the body's cells. Its "savings" consist mainly of glycogen and fat, which take longer to be brought out of store. Hunger is a sign that the body's energy reserves are running low, and that more food needs to be taken on board.

USING ENERGY

Most of the body's energy is produced by oxidizing or "burning" glucose, its chief fuel. Once this energy has been released, it is used to drive chemical reactions that keep cells alive, and it is also converted into kinetic (movement) energy, and heat energy. The body normally matches the energy that it takes in with the energy that it uses, so an Olympic sprinter needs more food than someone who has a less active way of life. But whatever a person's lifestyle, a healthy diet – combined with exercise – helps the body to keep its energy budget balanced.

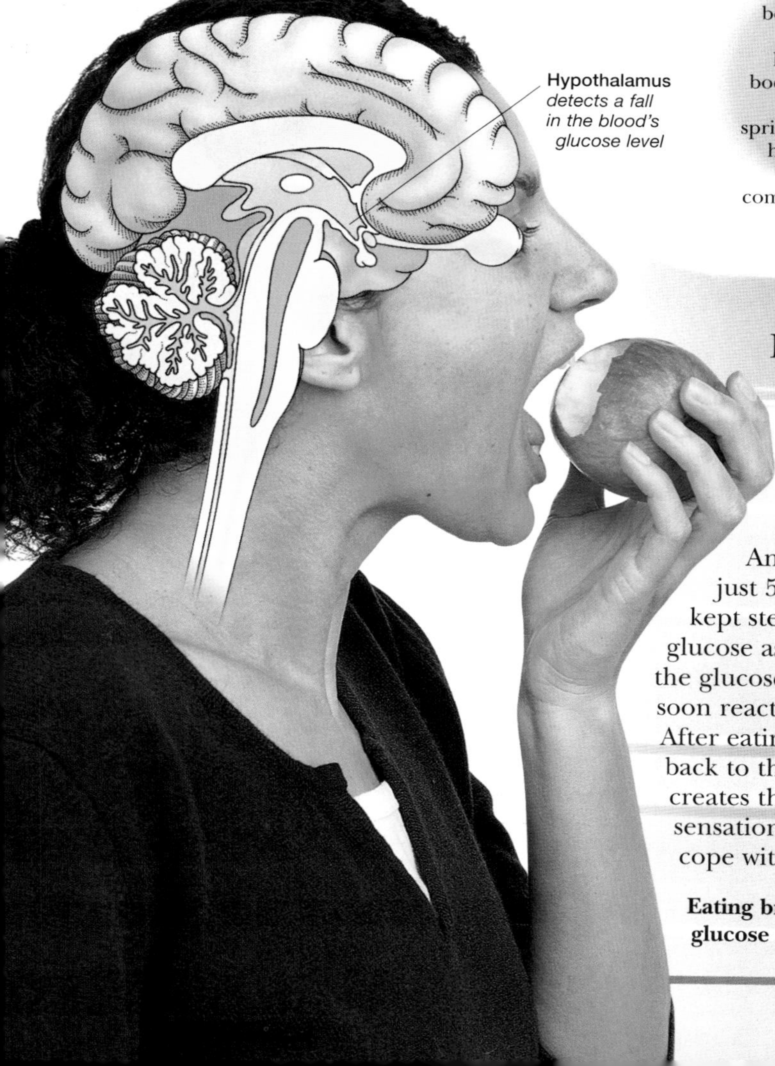

Hypothalamus *detects a fall in the blood's glucose level*

Feeling hungry

There is more to feeling hungry than having an empty stomach. Hunger is actually controlled by the hypothalamus – a region of the brain that monitors the concentration of glucose in the blood. An adult's bloodstream usually contains just 5 g (0.2 oz) of glucose, but the level is kept steady by the liver, which releases more glucose as existing stocks are used. However, if the glucose level starts to fall, the hypothalamus soon reacts, triggering the feeling of hunger. After eating, glucose from food then brings it back to the correct level. The hypothalamus also creates the feeling of fullness – an important sensation, because the digestive system cannot cope with too much food at one time.

Eating brings the blood's glucose level back to normal

Pizza
1,115 kJ (266 kcal)/ 100 g (3.5 oz). Pizzas are energy-rich because of their high fat content, but also provide some other nutrients.

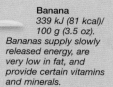

Banana
339 kJ (81 kcal)/ 100 g (3.5 oz). Bananas supply slowly released energy, are very low in fat, and provide certain vitamins and minerals.

ENERGY FROM FOOD

Foods contain different amounts of energy. For example, weight for weight, butter contains 15 times more energy than an apple. Traditionally, food energy is measured in calories (kcal), but most dieticians and food manufacturers now use the kilojoule (kJ) (1 kcal = 4.187 kJ). These four foods illustrate differences in energy content and show how, eaten in excess, some high-energy foods can unbalance the diet.

Salmon
837 kJ (200 kcal)/ 100 g (3.5 oz). Oily fish, like salmon, are rich in unsaturated fatty acids – "good" for the heart – proteins and certain vitamins.

Crisps
2,172 kJ (519 kcal)/100 g (3.5 oz). Crisps are very energy-rich because of their high fat content; they also contain a lot of salt.

A 55 kg (125 lb) cyclist travelling at 21 kph (13 mph) uses 2,220 kJ (534 kcal) during a one-hour ride

Obesity is a problem that affects not only adults, but also children who lead inactive lives

OUT OF BALANCE

The body uses fat as its long-term energy store, and as an "insurance policy" against hungry times. If a person takes in more food energy than they use, these fat stores slowly increase, and the person puts on weight. Reducing food intake and increasing activity levels will cause the body to dip into its fat reserves, and the weight begins to fall. Obesity is often caused by bad eating habits. However, weight loss can be a sign of disease, as well as a shortage of food.

ENERGY REQUIREMENTS

A person's daily energy requirements depend not only on whether they are male or female, but also on how old they are, and how active – or inactive – they are. Sex and age influence a person's basal metabolic rate (BMR), which is the rate at which the body releases energy when it is at rest. During adolescence, the body's energy need climbs steeply – not just because this is an active time of life, but also because the body is rapidly growing. Pregnant women, and mothers who breastfeed their babies, also have a high energy requirement, because they are literally "feeding for two".

AVERAGE DAILY ENERGY REQUIREMENTS

	kcal per day
Infant 9–12 months	891 kcal (3,730 kJ)
Child 5 years	1,627 kcal (6,810 kJ)
Child 8 years	1,853 kcal (7,760 kJ)
Child 11 years	2,029 kcal (8,495 kJ)
Girl 15 years	2,207 kcal (9,240 kJ)
Boy 15 years	2,875 kcal (12,035 kJ)
Woman (inactive)	1,917 kcal (8,025 kJ)
Woman (active)	2,150 kcal (9,000 kJ)
Woman (breastfeeding)	2,687 kcal (11,250 kJ)
Man (inactive)	2,515 kcal (10,530 kJ)
Man (active)	3,000 kcal (12,560 kJ)

kcal per day: 0   500   1,000   1,500   2,000   2,500   3,000

URINARY system

Several times each day, a person stops what they are doing in order to urinate. The urine they release is produced by the urinary system, a system that plays a key part in homeostasis – keeping conditions stable inside the body. At the core of the system are the two kidneys. Second by second, they process blood, removing two main components from it – unwanted wastes that must be excreted (eliminated) from the body before they build up and poison it, and water and salts that are surplus to requirements. The resulting watery waste – called urine – can then be expelled from the body.

EXCRETION

This is the elimination of chemical wastes from the body. Most of these wastes are unwanted products formed by metabolism in the body's cells, but some are substances taken in, but no longer required, by the body. The kidneys, for example, excrete urea, a nitrogenous (nitrogen-containing) waste produced in the liver. Other excretory processes include the loss of carbon dioxide from the lungs and sweat from the skin.

WASTE DISPOSAL SYSTEM

The major organs of the urinary system are the kidneys and the bladder. The kidneys filter the blood and produce urine, and the bladder stores urine until it can be expelled. The kidneys lie on either side of the back of the upper abdomen, whereas the bladder is found inside the pelvis. Urine is transported from the kidneys to the bladder along two tubes called the ureters, one from each kidney. A third tube called the urethra carries urine from the bladder so that it can leave the body during urination: in males the urethra runs along the length of the penis, and in females it comes to an end just in front of the opening of the vagina.

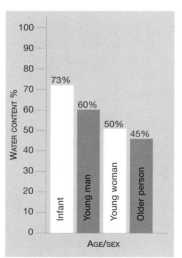

WATER CONTENT

Water is essential for the reactions that take place within cells, and it is the major component of the fluid that bathes the tissues of the body. Maintenance of constant water levels in the body is therefore essential, and this is one of the key functions of the kidneys. Factors such as age, sex, and fat content produce differences in the amount of water in different people's bodies, but the kidneys ensure that the water content of any one person remains constant.

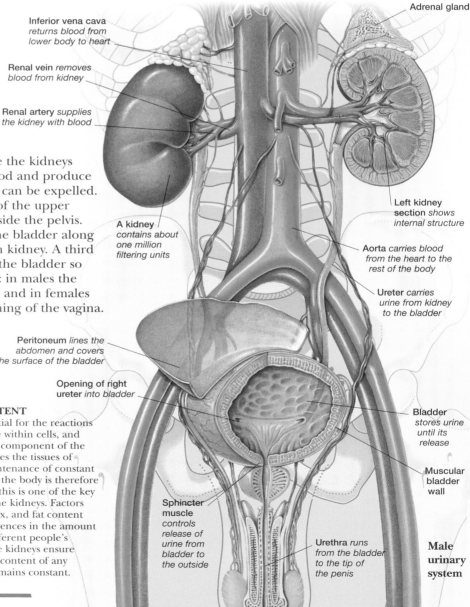

Inferior vena cava *returns blood from lower body to heart*

Renal vein *removes blood from kidney*

Renal artery *supplies the kidney with blood*

Adrenal gland

Left kidney section *shows internal structure*

A kidney *contains about one million filtering units*

Aorta *carries blood from the heart to the rest of the body*

Ureter *carries urine from kidney to the bladder*

Peritoneum *lines the abdomen and covers the surface of the bladder*

Opening of right ureter *into bladder*

Bladder *stores urine until its release*

Muscular bladder wall

Sphincter muscle *controls release of urine from bladder to the outside*

Urethra *runs from the bladder to the tip of the penis*

Male urinary system

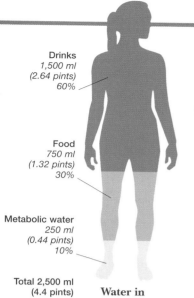

Drinks
1,500 ml
(2.64 pints)
60%

Food
750 ml
(1.32 pints)
30%

Metabolic water
250 ml
(0.44 pints)
10%

Total 2,500 ml
(4.4 pints)
Water in

Urine
1,500 ml
(2.64 pints)
60%

Losses via lungs
and skin 700 ml
(1.23 pints)
28%

Sweat
200 ml
(0.35 pints)
8%

Faeces
100 ml
(0.18 pints)
4%

Total 2,500 ml
(4.4 pints)
Water out

WATER BALANCE

The body matches the amount of water that leaves it to the amount that enters it, so that the total amount of water in the body remains constant. Water is obtained from food, drink, and the chemical reactions that take place within the body.
Loss of water occurs not just in the urine, but also in sweat and faeces, and in evaporation from the lungs and skin. The more water that enters the body, and the less that is lost in forms such as sweat, the more the kidneys must dispose of in the urine.

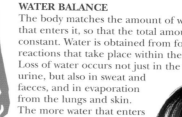

Drinking is a response to a signal issued by the thirst centre in the brain

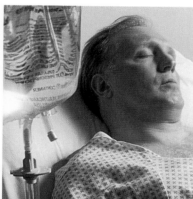

INTRAVENOUS INFUSION

Commonly known as a drip, an intravenous ("into a vein") infusion is used to introduce fluid directly into the bloodstream of patients who cannot maintain water balance. They may be unable to drink, or have lost large amounts of body fluid through vomiting, diarrhoea, or blood loss. The intravenous fluid passes downwards from a plastic container (above) into a vein in the arm.

CONTROLLING THIRST

Inside the brain's hypothalamus is a thirst centre which monitors the concentration of plasma – the liquid part of blood – passing through it. Plasma concentration increases as the body loses water. This leaves the mouth feeling dry because less saliva is produced, and the thirst centre generates a feeling of thirst. Having a drink soon quenches thirst, as plasma concentration returns to normal.

FUNCTIONS OF THE URINARY SYSTEM

Excretion	Removal of waste products and foreign chemicals from the blood.	Osmoregulation	Removal of excess salt and water ensures balance of concentration of body fluid.
Regulation of the blood pH	Hydrogen ions are removed from the blood and bicarbonate ions are added to control blood pH (acid-base balance).	Regulation of blood volume and composition	The various functions of the kidneys combine to maintain the correct volume and composition of the blood.
Regulation of blood pressure composition	Kidneys secrete renin which increases the amount of salt and water in the blood, and leads to constriction of blood vessels.	Other homeostatic roles	The kidneys release erythropoietin, a hormone that stimulates red blood cell production; and help synthesize the active form of vitamin D.

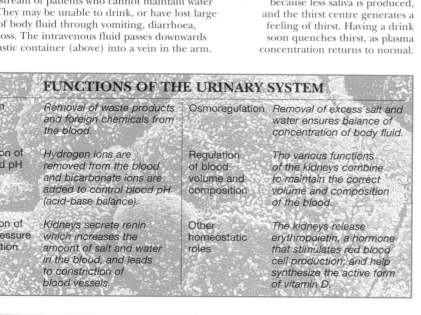

Kidneys

ATTACHED HIGH ENOUGH at the back of the abdominal cavity to be protected by the lowest ribs, the kidneys are perfectly adapted to their role of "cleansing" the blood and helping to keep its composition constant. The outer part of each kidney contains about one million tightly packed, coiled, tubular filtration units called nephrons. Each nephron receives a share of the copious supply of blood that is delivered every second by the renal artery and removed by the renal vein. Having filtered fluid from the blood, the nephron then processes the fluid to produce waste urine – as described in more detail on pp. 206–7. A continuous stream of urine trickles from the nephrons into the centre of the kidney, from where it is directed to the bladder.

KIDNEY STRUCTURE

About 12 cm (4.75 in) long and 3 cm (1.2 in) thick, each kidney is surrounded by a thin renal capsule. Internally, as this section (right) shows, the kidney is clearly divided into three zones: cortex, medulla, and pelvis. The outer cortex, and the medulla it surrounds, are the site of urine production. Fluid is filtered out of the blood in the cortex, while in the medulla, substances needed by the body are reabsorbed back from that fluid into the blood. The remaining urine is collected by ducts that open at the tips of the medulla's cone-shaped pyramids into the pelvis. This flattened, funnel-shaped tube channels urine into the ureter, which carries it to the bladder.

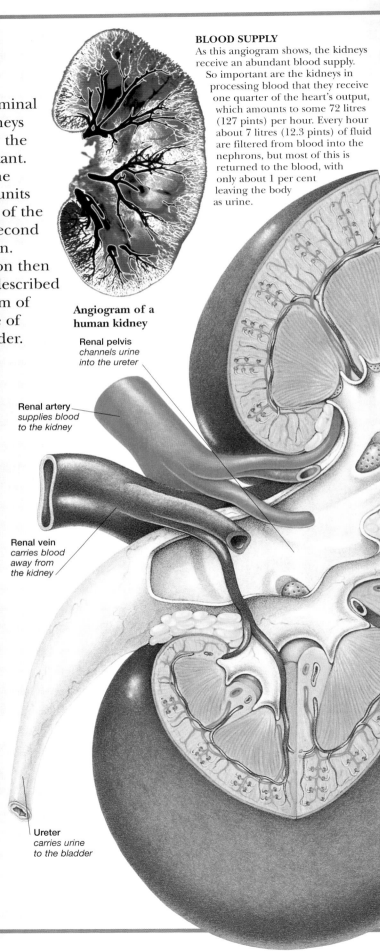

BLOOD SUPPLY
As this angiogram shows, the kidneys receive an abundant blood supply. So important are the kidneys in processing blood that they receive one quarter of the heart's output, which amounts to some 72 litres (127 pints) per hour. Every hour about 7 litres (12.3 pints) of fluid are filtered from blood into the nephrons, but most of this is returned to the blood, with only about 1 per cent leaving the body as urine.

Angiogram of a human kidney

Renal pelvis *channels urine into the ureter*

Renal artery *supplies blood to the kidney*

Renal vein *carries blood away from the kidney*

Ureter *carries urine to the bladder*

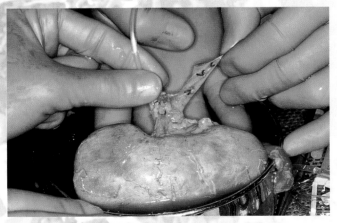

KIDNEY TRANSPLANTS
For people suffering from serious kidney disease, kidney transplantation is a life-saving treatment. The operation replaces a patient's defective kidney with a healthy one received from a donor. As with most forms of transplantation, drugs are required to prevent the person's immune system (see pp. 160–61) from rejecting the donor organ. Kidney transplantation is one of the most frequently performed types of transplantation.

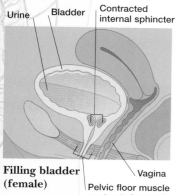

Filling bladder (female)

Labels: Urine · Bladder · Contracted internal sphincter · Vagina · Pelvic floor muscle *contracted around urethra*

Emptying bladder (female)

Labels: Bladder wall *contracts* · Relaxed internal sphincter · Urethra · Pelvic floor muscle *relaxed* · Urine *passes out through urethral opening*

URINATION

As these diagrams show, an internal sphincter – a ring of smooth muscle – at the junction of the bladder and urethra contracts to hold urine inside the filling bladder. A lower, external sphincter is provided by pelvic floor muscles contracting around the urethra. As the bladder fills, stretch receptors in its wall send nerve impulses to the spinal cord. In a reflex action, this returns signals telling the internal sphincter to relax and the bladder wall muscles to contract. At the same time, messages sent to the brain make a person feel the need to urinate. At a time of her choosing, she voluntarily relaxes the external sphincter, and bladder wall contractions push urine out.

UROSCOPY

In medieval times, doctors had few methods available to them for diagnosing diseases. Of those few, uroscopy, or "urine gazing", was the most common. There were up to 30 features of the patient's urine a doctor might note, including its colour, smell, cloudiness, and even its taste. These features were believed to provide information about the patient's health, thus enabling the doctor to prescribe a suitable course of treatment. In reality, uroscopy was of doubtful value, although it persisted until the 19th century. It was replaced then by urinology, the chemical analysis of urine for purposes of diagnosis, a practice that still continues today (see p. 207).

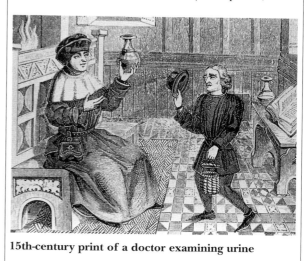

15th-century print of a doctor examining urine

INFANT NAPPIES

Until they reach about two years of age, children are unable to control the pelvic floor muscle – a skeletal (voluntary) muscle – that forms the external sphincter. As a result, as soon as the bladder has filled enough to trigger the spinal reflex, the internal sphincter opens, and the bladder contracts and empties through the external sphincter, which is already open. For that reason, babies, infants, and toddlers wear absorbent nappies until they have learned to control bladder emptying.

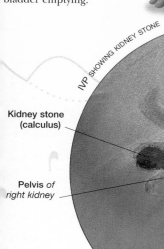

IVP SHOWING KIDNEY STONE

Labels: Kidney stone (calculus) · Pelvis *of right kidney* · Ureter

KIDNEY AND BLADDER STONES

Sometimes, certain wastes in urine can form crystals that remain and grow inside the kidney or bladder to make "stones". This intravenous pyelogram (IVP) shows a kidney stone lodged in the pelvis of the right kidney. Large stones remain where they are, but smaller ones can enter the ureter and cause excruciating pain. Bladder stones can also cause discomfort. A common remedy in the past was to remove stones by surgery, like the bladder stones below. But today, doctors use lithotripsy, a method that uses ultrasound to disintegrate stones into tiny fragments that can pass out of the body in the urine.

Bladder stones *that have been removed by surgery*

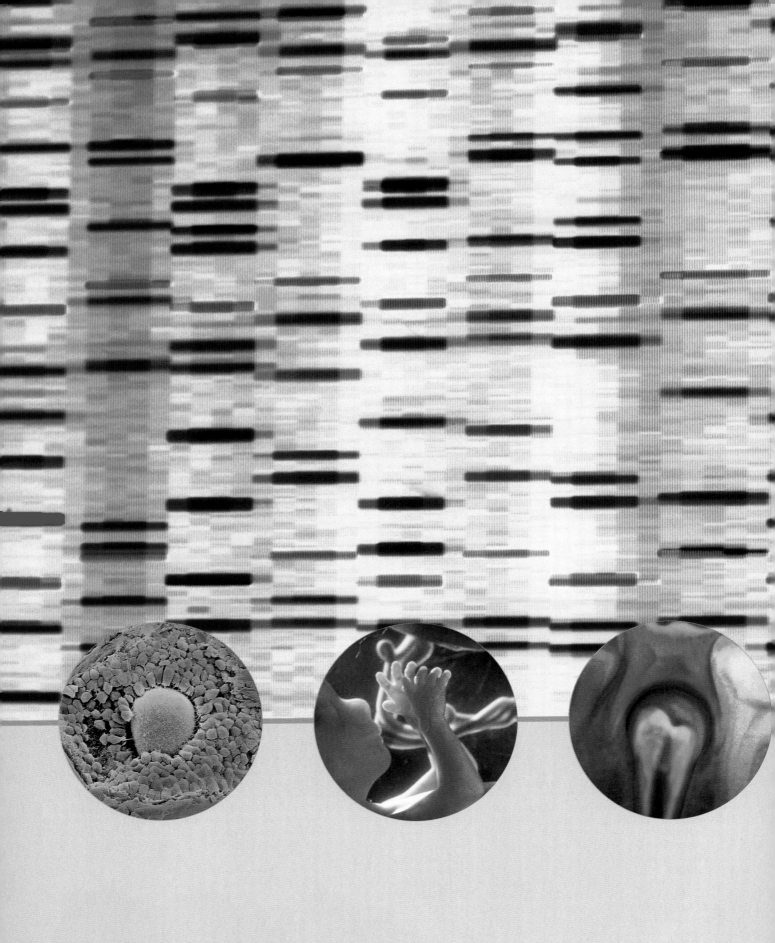

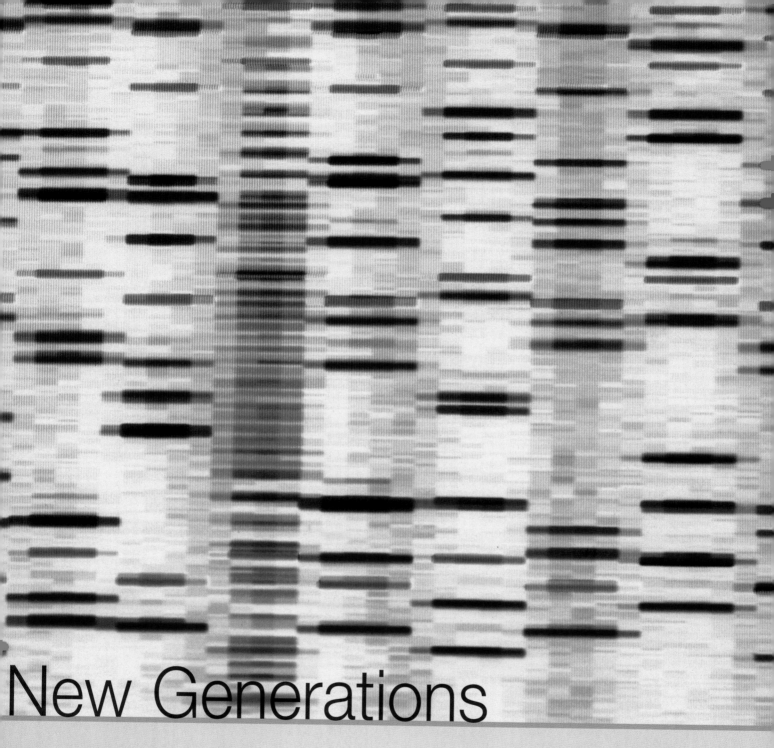

New Generations

EVERY PERSON ON EARTH follows the same life history that starts with their birth and ends with their death. During that life, many men and women reproduce, producing and nurturing a new generation of children who will eventually succeed them. Each child resembles their parents, but is not identical to either of them. These similarities and differences are controlled by thousands of chemical instructions called genes found inside every body cell. The identity and structure of these genes is currently under investigation.

REPRODUCTIVE systems

REPRODUCTION IS A complex process in which genetic material from two parents is brought together to form a new human being. This material is contained within special sex cells – sperm, which are produced by males, and ova (or eggs), produced by females. In order for reproduction to occur, a sperm and an ovum must join to form a single cell containing a full set of genetic material, half each from the father and mother. The reproductive process continues over nine months of pregnancy, when this single cell develops into a fully formed baby, culminating in childbirth.

Female reproductive system

Male reproductive system

Ovaries *produce ova*

Uterus, *where the fetus develops*

Fallopian tubes *lead from the ovaries to the uterus*

Vagina *receives the penis during sexual intercourse*

Penis *delivers sperm inside the female*

Testes *produce sperm*

REPRODUCTIVE SYSTEMS

Males and females each have specialized reproductive organs. In the male these are the testes and the penis, while the female organs are the ovaries, the fallopian tubes, the uterus, and the vagina. The reproductive systems of both sexes produce sex cells, and enable these cells to meet during sexual intercourse so that fertilization can occur. The female system has a further function not found in the male – it provides an environment in which conception and fetal development occur, and from which the baby emerges during birth.

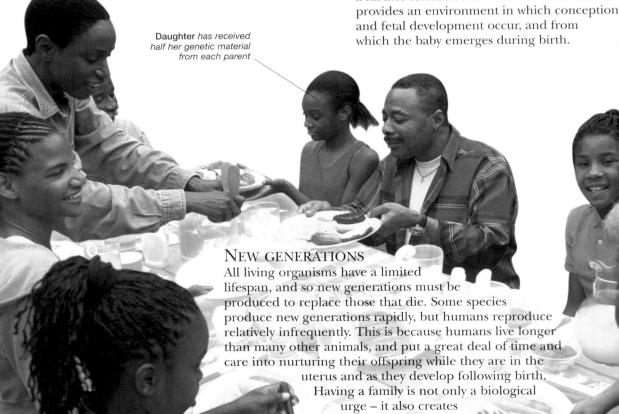

Daughter *has received half her genetic material from each parent*

NEW GENERATIONS

All living organisms have a limited lifespan, and so new generations must be produced to replace those that die. Some species produce new generations rapidly, but humans reproduce relatively infrequently. This is because humans live longer than many other animals, and put a great deal of time and care into nurturing their offspring while they are in the uterus and as they develop following birth. Having a family is not only a biological urge – it also creates strong social bonds of support and love.

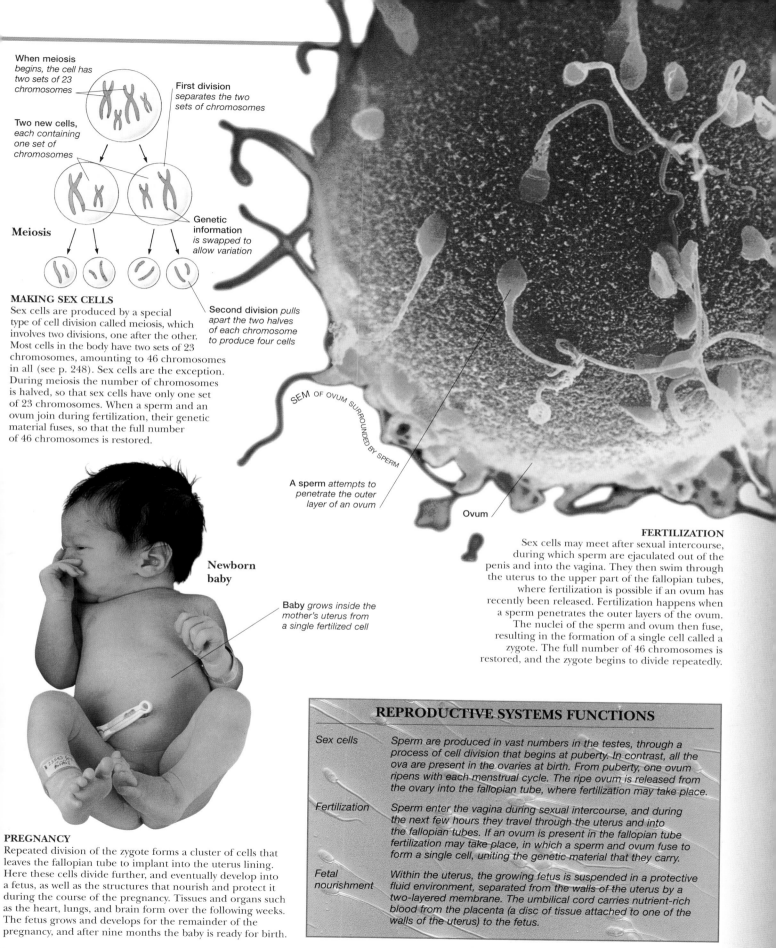

When meiosis *begins, the cell has two sets of 23 chromosomes*

First division *separates the two sets of chromosomes*

Two new cells, *each containing one set of chromosomes*

Meiosis

Genetic information *is swapped to allow variation*

MAKING SEX CELLS

Sex cells are produced by a special type of cell division called meiosis, which involves two divisions, one after the other. Most cells in the body have two sets of 23 chromosomes, amounting to 46 chromosomes in all (see p. 248). Sex cells are the exception. During meiosis the number of chromosomes is halved, so that sex cells have only one set of 23 chromosomes. When a sperm and an ovum join during fertilization, their genetic material fuses, so that the full number of 46 chromosomes is restored.

Second division *pulls apart the two halves of each chromosome to produce four cells*

SEM OF OVUM SURROUNDED BY SPERM

A sperm *attempts to penetrate the outer layer of an ovum*

Ovum

Newborn baby

Baby *grows inside the mother's uterus from a single fertilized cell*

FERTILIZATION

Sex cells may meet after sexual intercourse, during which sperm are ejaculated out of the penis and into the vagina. They then swim through the uterus to the upper part of the fallopian tubes, where fertilization is possible if an ovum has recently been released. Fertilization happens when a sperm penetrates the outer layers of the ovum. The nuclei of the sperm and ovum then fuse, resulting in the formation of a single cell called a zygote. The full number of 46 chromosomes is restored, and the zygote begins to divide repeatedly.

PREGNANCY

Repeated division of the zygote forms a cluster of cells that leaves the fallopian tube to implant into the uterus lining. Here these cells divide further, and eventually develop into a fetus, as well as the structures that nourish and protect it during the course of the pregnancy. Tissues and organs such as the heart, lungs, and brain form over the following weeks. The fetus grows and develops for the remainder of the pregnancy, and after nine months the baby is ready for birth.

REPRODUCTIVE SYSTEMS FUNCTIONS

Sex cells	Sperm are produced in vast numbers in the testes, through a process of cell division that begins at puberty. In contrast, all the ova are present in the ovaries at birth. From puberty, one ovum ripens with each menstrual cycle. The ripe ovum is released from the ovary into the fallopian tube, where fertilization may take place.
Fertilization	Sperm enter the vagina during sexual intercourse, and during the next few hours they travel through the uterus and into the fallopian tubes. If an ovum is present in the fallopian tube fertilization may take place, in which a sperm and ovum fuse to form a single cell, uniting the genetic material that they carry.
Fetal nourishment	Within the uterus, the growing fetus is suspended in a protective fluid environment, separated from the walls of the uterus by a two-layered membrane. The umbilical cord carries nutrient-rich blood from the placenta (a disc of tissue attached to one of the walls of the uterus) to the fetus.

Male reproductive system

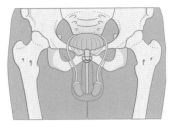

Male reproductive system

THE ORGANS OF A MAN'S reproductive system enable him to make sperm, have sexual intercourse, and fertilize ova produced by a female. The outer, visible parts of the male reproductive system are the penis and a pouch called the scrotum. Inside the scrotum is a pair of oval glands called testes, which are the main reproductive organs of the male. The testes produce millions of minute sex cells, called sperm, and also make male sex hormones. During sexual intercourse, sperm are transferred into the female's vagina by the penis. The sperm are carried from the testes through a series of internal tubes. They are released from the body in a fluid called semen, which is formed from liquids secreted by the seminal vesicles and the prostate gland.

REPRODUCTIVE ORGANS

The testes – the sperm-makers – hang outside the body, inside the scrotum and below the penis. A complicated tube system delivers sperm from the testes to the penis. Small tubes lead from each testis into the long, tightly coiled epididymis, located behind the testis. Each epididymis leads into another tube, called the vas deferens, which joins with a small tube from one of the seminal vesicles to form the ejaculatory duct. The two ejaculatory ducts then join the urethra as it passes through the prostate gland. The urethra is the channel that passes from the bladder to the tip of the penis, through which sperm are released.

PENIS

The penis contains three columns of spongy tissue. During sexual arousal, the spongy columns fill with blood to make the penis larger, firm, and erect, ready for sexual intercourse. Sperm is ejaculated through the urethra, a tube that passes through the middle of the penis to its tip. Urine from the bladder also passes out of the body through the urethra. The tip of the penis is covered by the foreskin, which may be removed during an operation called circumcision.

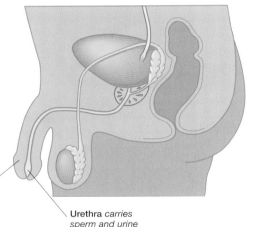

Vas deferens *is a tube that carries sperm from a testis to an ejaculatory duct*

Penis *delivers sperm into the female's vagina during sexual intercourse*

Urethra *is a tube that carries sperm and urine to the outside of the body*

Corpora cavernosa *are two parallel cylinders of tissue that fill with blood to make the penis erect*

Corpus spongiosum *is spongy tissue that surrounds the urethra and fills with blood to make the penis erect*

Foreskin *covers and protects the head of the penis*

Glans penis *is the head of the penis*

Testes *produce sperm and sex hormones*

Penis *becomes erect when blood fills the spongy tissue inside*

Urethra *carries sperm and urine out of the body*

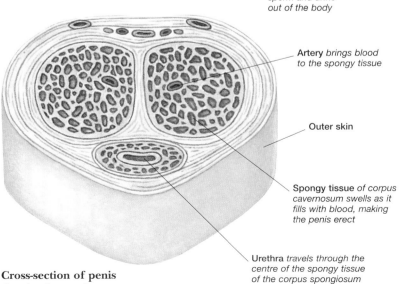

Artery *brings blood to the spongy tissue*

Outer skin

Spongy tissue *of corpus cavernosum swells as it fills with blood, making the penis erect*

Urethra *travels through the centre of the spongy tissue of the corpus spongiosum*

Cross-section of penis showing internal structure

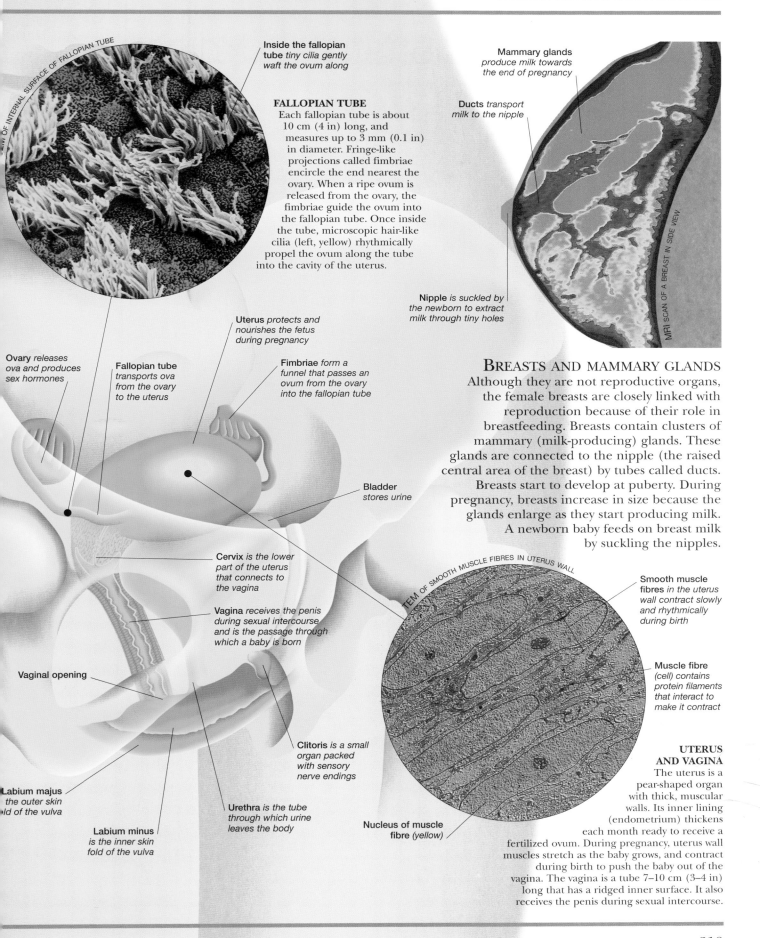

EM OF INTERNAL SURFACE OF FALLOPIAN TUBE

Inside the fallopian tube *tiny cilia gently waft the ovum along*

FALLOPIAN TUBE
Each fallopian tube is about 10 cm (4 in) long, and measures up to 3 mm (0.1 in) in diameter. Fringe-like projections called fimbriae encircle the end nearest the ovary. When a ripe ovum is released from the ovary, the fimbriae guide the ovum into the fallopian tube. Once inside the tube, microscopic hair-like cilia (left, yellow) rhythmically propel the ovum along the tube into the cavity of the uterus.

Mammary glands *produce milk towards the end of pregnancy*

Ducts *transport milk to the nipple*

MRI SCAN OF A BREAST IN SIDE VIEW

Nipple *is suckled by the newborn to extract milk through tiny holes*

Uterus *protects and nourishes the fetus during pregnancy*

Ovary *releases ova and produces sex hormones*

Fallopian tube *transports ova from the ovary to the uterus*

Fimbriae *form a funnel that passes an ovum from the ovary into the fallopian tube*

BREASTS AND MAMMARY GLANDS
Although they are not reproductive organs, the female breasts are closely linked with reproduction because of their role in breastfeeding. Breasts contain clusters of mammary (milk-producing) glands. These glands are connected to the nipple (the raised central area of the breast) by tubes called ducts. Breasts start to develop at puberty. During pregnancy, breasts increase in size because the glands enlarge as they start producing milk. A newborn baby feeds on breast milk by suckling the nipples.

Bladder *stores urine*

Cervix *is the lower part of the uterus that connects to the vagina*

Vagina *receives the penis during sexual intercourse and is the passage through which a baby is born*

Vaginal opening

TEM OF SMOOTH MUSCLE FIBRES IN UTERUS WALL

Smooth muscle fibres *in the uterus wall contract slowly and rhythmically during birth*

Muscle fibre *(cell) contains protein filaments that interact to make it contract*

Clitoris *is a small organ packed with sensory nerve endings*

Labium majus *the outer skin ...ld of the vulva*

Labium minus *is the inner skin fold of the vulva*

Urethra *is the tube through which urine leaves the body*

Nucleus of muscle fibre *(yellow)*

UTERUS AND VAGINA
The uterus is a pear-shaped organ with thick, muscular walls. Its inner lining (endometrium) thickens each month ready to receive a fertilized ovum. During pregnancy, uterus wall muscles stretch as the baby grows, and contract during birth to push the baby out of the vagina. The vagina is a tube 7–10 cm (3–4 in) long that has a ridged inner surface. It also receives the penis during sexual intercourse.

Ovarian and menstrual cycles

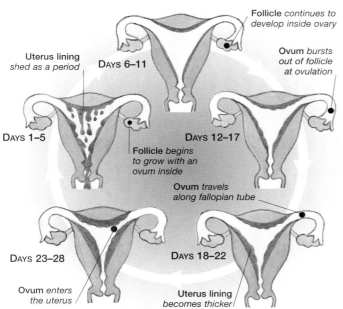

Follicle *continues to develop inside ovary*

DAYS 6–11

Uterus lining *shed as a period*

Ovum *bursts out of follicle at ovulation*

DAYS 1–5

DAYS 12–17

Follicle *begins to grow with an ovum inside*

Ovum *travels along fallopian tube*

DAYS 23–28

DAYS 18–22

Ovum *enters the uterus*

Uterus lining *becomes thicker*

EACH MONTH, A WOMAN'S REPRODUCTIVE SYSTEM goes through a set sequence of changes – a cycle – that prepares her body for the possibility of pregnancy. This reproductive cycle has two interlinked parts – the ovarian cycle and the menstrual cycle. During the ovarian cycle, an ovum (egg) ripens inside one of the ovaries and is then released. During the menstrual cycle, the lining of the uterus (the endometrium) thickens in readiness to receive the ovum if it is fertilized. If fertilization does not happen, the endometrium breaks down and is shed through the vagina during menstruation (a period). The monthly changes in the ovaries and the uterus are controlled by hormones from the brain and the ovaries.

REPRODUCTIVE CYCLES

The menstrual and ovarian cycles usually begin when a girl is between 11 and 14 years of age and stop when a woman reaches the menopause (at about 50). The cycles repeat every 28 days on average, but this varies between women and can be different each month. From puberty until the menopause, the endometrium is repeatedly shed during a period and then thickens again, while a follicle in the ovary matures and releases its ovum into the fallopian tube. If the ovum reaches the uterus unfertilized, it is shed with the endometrium, marking the end of one cycle and the beginning of the next.

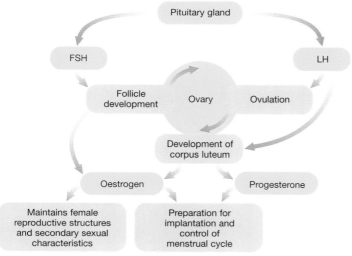

Pituitary gland

FSH

LH

Follicle development

Ovary

Ovulation

Development of corpus luteum

Oestrogen

Progesterone

Maintains female reproductive structures and secondary sexual characteristics

Preparation for implantation and control of menstrual cycle

SEX HORMONES

Four hormones control the female reproductive system: FSH (follicle-stimulating hormone) and LH (luteinizing hormone) from the pituitary gland; and the sex hormones oestrogen and progesterone from the ovaries. After puberty, FSH and LH stimulate follicle maturation and ovulation. Ripening follicles release oestrogen, which makes the endometrium thicken and maintains the female sexual organs. After ovulation, progesterone is released which also has a controlling role in the menstrual cycle.

OVARIAN CYCLE

At the start of each ovarian cycle, several follicles – each containing an ovum – begin to enlarge and fill up with fluid. One follicle becomes fully mature and bursts, releasing its ovum from the surface of the ovary during ovulation. The empty follicle forms the corpus luteum, a mass of cells that release hormones.

MENSTRUAL CYCLE

At the start of each cycle, the endometrium is shed during menstruation. The bleeding lasts an average of five days, and contains the unfertilized ovum as well as cells from the uterine lining. Hormones are then released that cause the endometrium to increase in size to a thickness of 6 mm (0.25 in) before being shed again.

HORMONE LEVELS

Hormone levels fluctuate over the month because of the interaction between them. Initially, FSH causes an ovum to mature, and its follicle releases oestrogen. Increasing oestrogen inhibits FSH, but stimulates LH, which surges, stimulating the ripe follicle to burst and release its ovum. The newly formed corpus luteum then releases progesterone, which further stimulates endometrial thickening, and some oestrogen. Both hormones suppress FSH and LH. Without fertilization, the corpus luteum breaks down, progesterone and oestrogen levels fall, and the cycle begins again.

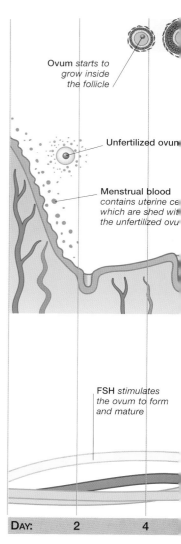

Ovum *starts to grow inside the follicle*

Unfertilized ovum

Menstrual blood *contains uterine ce which are shed wit the unfertilized ovu*

FSH *stimulates the ovum to form and mature*

DAY: 2 4

OVULATION

About halfway through the ovarian (and menstrual) cycle, ovulation occurs. One follicle outgrows the others and matures, producing a blister-like swelling just under the surface of the ovary. The follicle eventually bursts, and the ovum is released from the ovary. It is then moved along the fallopian tube, where it may be fertilized, and into the uterus.

SEM OF AN OVUM AT OVULATION

Uterus lining *has folds and a spongy surface after ovulation*

SEM OF ENDOMETRIUM

Ovum *(pink) bursts from the ovary's surface*

SPONGY SURFACE

Both before and after ovulation, hormones from the ovaries cause the endometrium to thicken and its blood supply to increase. As the tissue gets thicker and starts to look spongy, it is ready to receive a fertilized ovum if one embeds itself into the tissue. Glands in the surface (darker areas) secrete nutrients, such as lipids and sugars, to nourish the fertilized ovum.

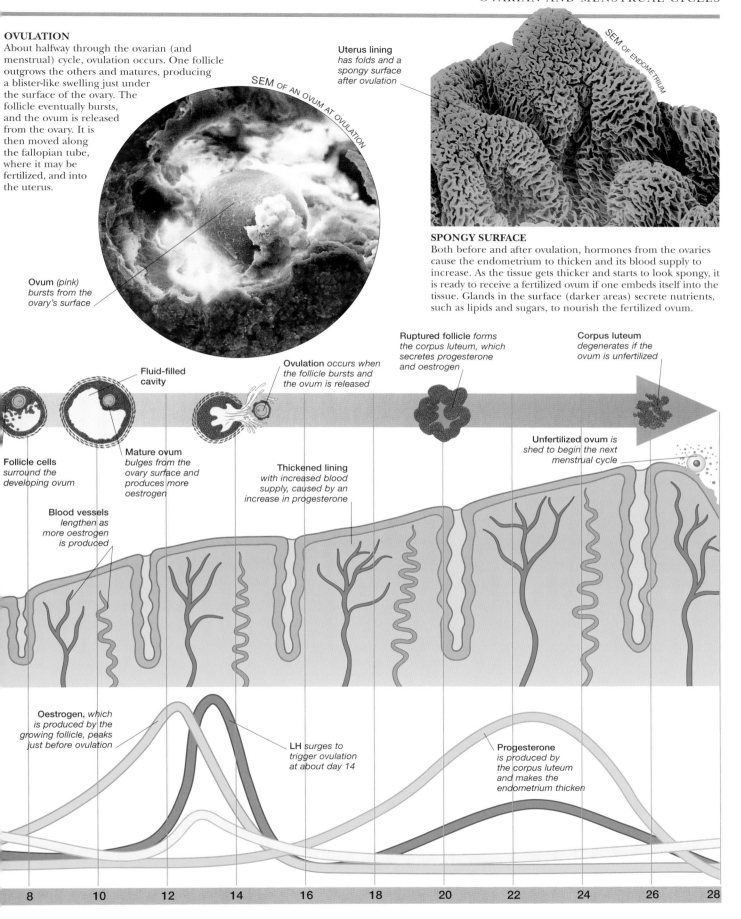

Ruptured follicle *forms the corpus luteum, which secretes progesterone and oestrogen*

Corpus luteum *degenerates if the ovum is unfertilized*

Fluid-filled cavity

Ovulation *occurs when the follicle bursts and the ovum is released*

Unfertilized ovum *is shed to begin the next menstrual cycle*

Follicle cells *surround the developing ovum*

Mature ovum *bulges from the ovary surface and produces more oestrogen*

Thickened lining *with increased blood supply, caused by an increase in progesterone*

Blood vessels *lengthen as more oestrogen is produced*

Oestrogen, *which is produced by the growing follicle, peaks just before ovulation*

LH *surges to trigger ovulation at about day 14*

Progesterone *is produced by the corpus luteum and makes the endometrium thicken*

| 8 | 10 | 12 | 14 | 16 | 18 | 20 | 22 | 24 | 26 | 28 |

Conception

THE CREATION OF EACH NEW HUMAN LIFE starts with fertilization. This is the union of an ovum (egg) from a female with a sperm from a male. The male and female cells are brought together following sexual intercourse, when the male's erect penis is placed inside the female's vagina. During intercourse, the male releases semen through the opening at the tip of the penis into the woman's vagina. The millions of sperm in the semen swim through the cervix into the uterus and towards the ovum in the fallopian tube, although only a few actually reach the ovum. Fertilization usually takes place inside a fallopian tube, but it can only occur if sperm meets the ovum in the first 24 hours after the ovum's release from the ovary. Once a sperm penetrates the ovum, its nucleus fuses with that of the ovum and a single new cell (zygote) is produced that contains genetic material from both parents. The new cell divides repeatedly to form a ball-shaped cluster of cells, which travels along the fallopian tube into the uterus. About a week after fertilization, the tiny ball of cells burrows, or implants, into the lining of the uterus. The start of pregnancy, which begins with fertilization and implantation, is called conception.

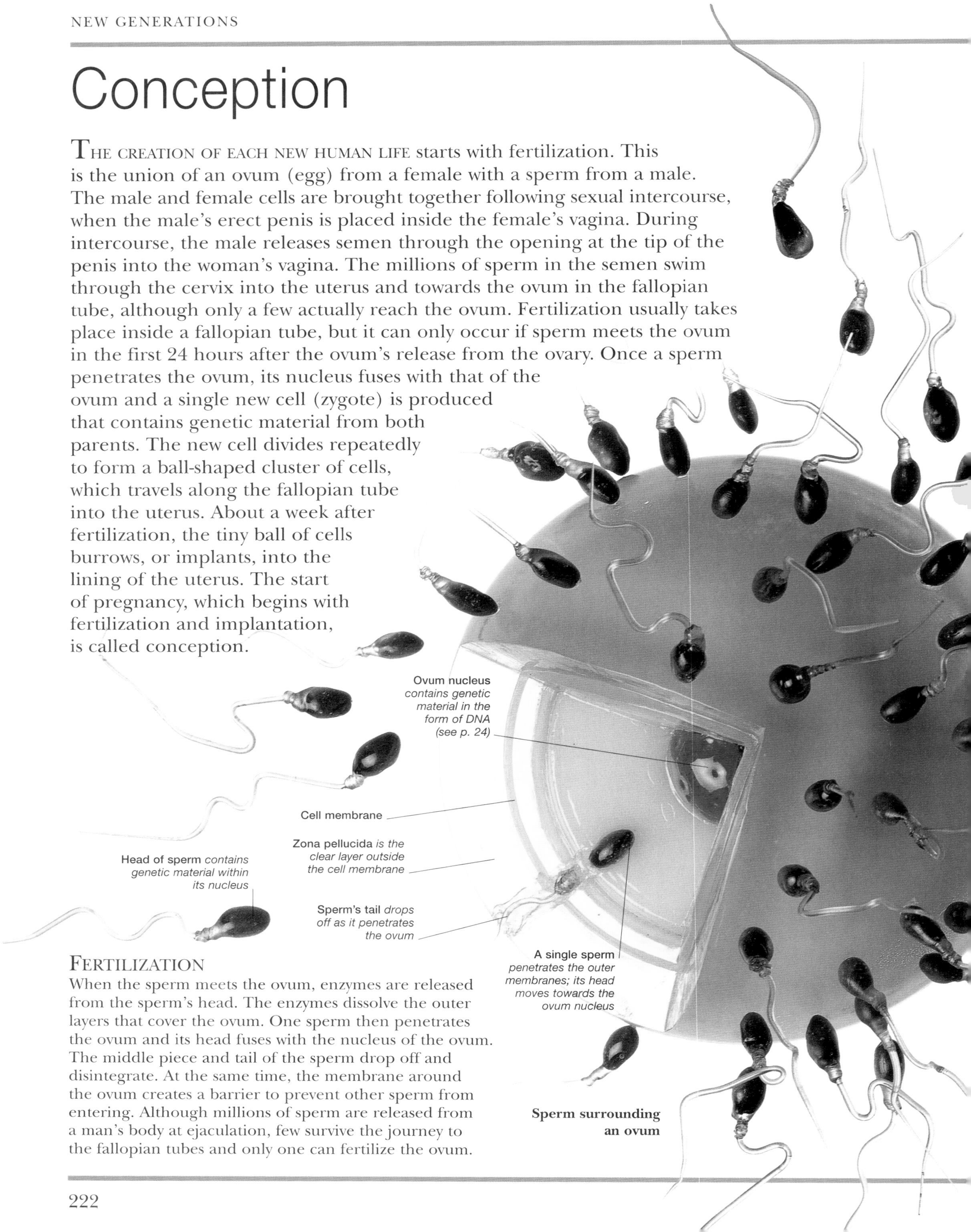

Ovum nucleus *contains genetic material in the form of DNA (see p. 24)*

Cell membrane

Zona pellucida *is the clear layer outside the cell membrane*

Head of sperm *contains genetic material within its nucleus*

Sperm's tail *drops off as it penetrates the ovum*

A single sperm *penetrates the outer membranes; its head moves towards the ovum nucleus*

FERTILIZATION

When the sperm meets the ovum, enzymes are released from the sperm's head. The enzymes dissolve the outer layers that cover the ovum. One sperm then penetrates the ovum and its head fuses with the nucleus of the ovum. The middle piece and tail of the sperm drop off and disintegrate. At the same time, the membrane around the ovum creates a barrier to prevent other sperm from entering. Although millions of sperm are released from a man's body at ejaculation, few survive the journey to the fallopian tubes and only one can fertilize the ovum.

Sperm surrounding an ovum

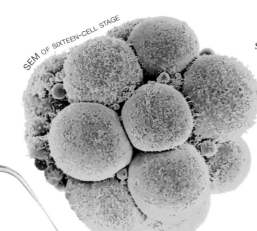

SEM OF TWO-CELL STAGE

TWO-CELL STAGE – 36 HOURS
It takes about 36 hours for the fertilized cell to divide into two (above) as it continues along the fallopian tube. Cell division by mitosis (see p. 19) creates two identical new cells, each with an exact copy of all the genetic information contained in the original cell. At this stage, the zygote measures about 0.1 mm (0.004 in) across.

SEM OF FOUR-CELL STAGE

FOUR-CELL STAGE – 48 HOURS
As the zygote travels further along the fallopian tube, it continues to divide. Here, after about 48 hours, the cells divide again to become a four-cell ball. As its journey continues, the cells divide every 12 hours or so.

SEM OF SIXTEEN-CELL STAGE

SIXTEEN-CELL STAGE – 72 HOURS
Three days after fertilization, the cells create a sixteen-cell, solid, berry-like ball called a morula (left). This stage gets its name from the Latin word for "mulberry". The morula eventually finishes its journey down the fallopian tube and enters the uterus. It then changes into a hollow ball of cells called a blastocyst.

IMPLANTATION – SIX DAYS
Six days after fertilization, the blastocyst (orange) implants itself into the lining of the uterus (pink), marking the end of conception. Its inner cells become the embryo, while the outer cells help form the placenta. The placenta connects the growing embryo to its mother and nourishes it. After day 10, the embryo is completely embedded and continues its growth.

EARLY DAYS
After fertilization, the zygote divides into two, then again into four, and so on. Muscular contractions combine with beating cilia to gently waft the growing ball of cells along the fallopian tube. After about three days, the dividing cells have formed a berry-like cluster called a morula, which reaches the end of the fallopian tube and enters the uterus. About a week after fertilization, the morula develops into a fluid-filled ball of cells called a blastocyst and implants into the uterus lining, beginning pregnancy.

Fertilization occurs as a sperm penetrates the ovum, creating a zygote

Two-cell stage occurs as the zygote begins to divide and grow

Fallopian tube

Four-cell stage

Ovulation is when an egg bursts from the ovary and moves along the fallopian tube

Ovary

16-cell stage is a cluster of dividing cells called a morula

Uterus

Implantation occurs when the blastocyst reaches the uterus and embeds itself into the lining

Journey of the fertilized ovum

SEM OF IMPLANTED EMBRYO

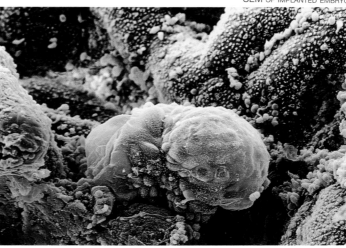

FERTILITY

MODERN MEDICINE HAS DEVELOPED ways to control human fertility – both to prevent it and aid it. The fact that women can produce large numbers of children over 30 or 40 fertile years, and the need to give each child a long period of parental care, has led to the search for methods of contraception to control the number of births and the length of time between them. Conversely, some people have difficulty conceiving children due to problems in their reproductive systems. Modern fertility treatments have made it possible for many childless couples to have a family, by joining ova and sperm in a laboratory.

DR MARIE STOPES
Dr Marie Stopes campaigned for the right for women to use birth control, publishing two bestselling books on the subject. In 1921, she opened the first British birth control clinic.

BIRTH CONTROL PILL
There are two main types of birth control pill, the combined pill and the progestogen-only pill. The hormones in them usually prevent a woman from ovulating. This means that no ova are released, so they cannot be fertilized by sperm.

BARRIER METHODS
The earliest attempts at contraception centred on the use of physical barriers that stopped sperm from reaching the ovum, for example condoms that cover the penis. Their widespread availability today, both as a contraceptive and a barrier to sexually transmitted diseases, was first pioneered by Margaret Sanger (1879–1966) in America and Marie Stopes (1880–1958) in Britain. They both overcame strong social and religious opposition to make contraception freely available, so that women could choose when they would become pregnant.

IUD AND BIRTH CONTROL PILLS
More sophisticated methods of contraception, based on treatments which interfere with the processes of ovum production and fertilization,

became available in the second half of the 20th century. Small interuterine devices (IUDs), implanted inside the uterus, work by preventing fertilized ova from implanting. More recent interuterine systems (IUS) contain the hormone progestogen, which stops sperm entering the uterus.

The use of hormones to regulate ovulation was pioneered by several scientists, whose studies explained how hormones controlled human fertility. They developed artificial hormones that stop ova maturing, and these formed the basis of birth control pills, taken orally. Their widespread introduction in the 1960s meant that women could enjoy sexual intercourse without the fear of unwanted pregnancy. Although

Fallopian tube
is blocked by a clip during sterilization

FEMALE STERILIZATION
This method of contraception is permanent and involves surgery. Two small incisions are made into the abdomen. The fallopian tubes are then sealed with clips or cut and tied, so that sperm cannot travel through the tubes to fertilize ova. Males can be sterilized by cutting the vas deferens.

FINAL STAGE

The final stage of labour begins as soon as the baby has been born, and lasts until the placenta is delivered. Mild contractions continue after the birth of the baby, and the placenta gently peels away from the inner surface of the uterus and follows the umbilical cord out through the vagina. Blood vessels in the wall of the uterus that previously supplied the placenta clamp shut, preventing excessive bleeding. The placenta, or afterbirth, is usually delivered about 5–30 minutes after the birth of the baby, and consists of a disc of spongy, blood-rich tissue about 20 cm (8 in) in diameter.

Placenta *peels away from the uterus lining*

Cut umbilical cord

APGAR SCORE

At birth, and during the following minutes, the baby's well-being can be assessed using a tool called the Apgar score, developed by an American doctor, Virginia Apgar, in which vital signs are evaluated. A low Apgar score indicates that the baby is unwell.

SIGN	SCORE: 0	SCORE: 1	SCORE: 2
Activity	Limp	Some bending of limbs	Active movements
Pulse	None	Below 100 bpm	Over 100 bpm
Grimace	None	Grimace or whimpering	Cry, sneeze, or cough
Appearance	Pale, blue	Blue extremities	Pink
Respiration	None	Slow or irregular breaths, weak cry	Regular breaths, strong cry

Vernix *is a greasy substance that covers and protects the baby's skin until it is born*

Umbilical cord *is clamped and cut once blood has flowed back to the baby*

BREASTFEEDING

During pregnancy, hormones activate the milk-producing glands of the breasts, so that food is available for the newborn. At first, the breasts produce colostrum, a yellowish fluid that is rich in antibodies, which protect the baby from infection. During the second day after birth, colostrum is replaced by ordinary breast milk, which contains lactose (a type of sugar), protein, and fat. Usually, a mother produces about 1 litre (1.76 pints) of milk a day for as long as she wishes to breastfeed.

SAVING MOTHERS

THE BIRTH OF A BABY SHOULD be a time of happiness for parents. Yet, although the process is natural, it is never risk-free and, tragically, a few mothers die during or following childbirth. The number of fatalities today is tiny when compared with earlier centuries. In the 18th and 19th centuries, for example, many women died within a week or two of giving birth from a disease called puerperal (childbed) fever, particularly in hospitals. In Europe and America, a handful of doctors realized both how puerperal fever might be spread and be prevented. But their findings were ignored, and it took decades before measures were taken to save mothers.

WASHING HANDS
To many 19th-century doctors, the idea that washing their hands would stop the spread of disease seemed ridiculous. Today, hand washing is a vital part of stopping infection passing from doctor or midwife to patient.

IGNAZ SEMMELWEIS
Deaths of women from puerperal fever in Vienna General Hospital were greatly reduced by the work of Ignaz Semmelweis.

THE MENACE OF INFECTION
By the end of the 18th century there had been great advances in the understanding of pregnancy and childbirth, with the training of specialist doctors, called obstetricians, and the opening of maternity hospitals. Although these hospitals provided bed rest for poorer expectant mothers, they also had appalling death rates. Within days of giving birth, many mothers died from puerperal

EARLY MATERNITY WARD
This illustration, taken from the book *Microcosm of London* published in 18 shows the scene inside a woman's ward in the Middlesex Hospital, London, England. The view was probably idealized by the artist, as conditions were certainly more crowded and dirty than shown here.

Speech and vocabulary *becomes complex as childhood progresses*

SPEECH
Speech develops at varying rates in young children. Gradually actual words replace the "babble" of infancy. From about the age of two the child begins to combine groups of words to convey more complex ideas, and by four he or she is usually able to engage in simple conversation. During the remaining years of childhood, speech rapidly increases in complexity as vocabulary and grammatical understanding develop.

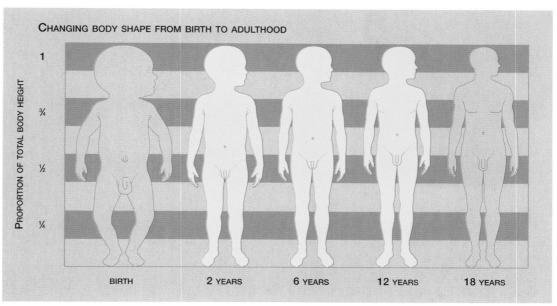

CHANGING BODY SHAPE FROM BIRTH TO ADULTHOOD

PROPORTION OF TOTAL BODY HEIGHT

1 ¾ ½ ¼

BIRTH   2 YEARS   6 YEARS   12 YEARS   18 YEARS

BODY PROPORTIONS
Children grow rapidly during the first two years of life. Growth then continues at a reduced rate for the remainder of childhood – usually about 7 cm (2.75 in) in height and 2.7 kg (6 lb) in weight each year. Different parts of the body grow at varying rates. In a young child the head accounts for a large part of the total height of the body, but it becomes less prominent as the limbs and trunk grow longer. Over time, the proportions of the body gradually begin to resemble those of an adult.

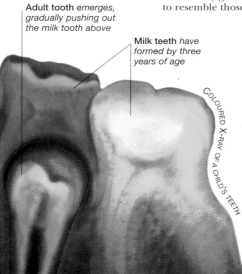

Adult tooth *emerges, gradually pushing out the milk tooth above*

Milk teeth *have formed by three years of age*

COLOURED X-RAY OF A CHILD'S TEETH

COLOURED X-RAY OF A CHILD'S SKULL

FACE SHAPE
At birth and throughout infancy, copious reserves of fat beneath the skin give the face a rounded appearance. This persists well into childhood, so that during the first years of life the finer features of the face have yet to become apparent. At about the age of five, this begins to change – the fat reserves of babyhood ebb away, and the face becomes leaner and the features more distinct. Family resemblances that may previously have been vague often become more pronounced at this time. Facial bones become larger and stronger as childhood passes, also changing the face shape.

TEETH
By the end of infancy, a child usually has two front teeth in the upper jaw and two in the lower. More teeth emerge during the following months, so that the full set of 20 milk teeth (deciduous teeth) is in place by about the age of three. Teething can be painful, so while a tooth is emerging the child may be irritable and tearful. At about the age of six, permanent teeth begin to emerge, and as they do so, each dislodges the overlying milk tooth, which falls out. Permanent teeth have usually replaced milk teeth by adolescence (see pp. 180–1).

Adolescence

ADOLESCENCE IS A TIME OF transition from childhood to adulthood, and involves both physical and psychological changes. Through a process called puberty, the sexual organs become functional. As a result, sex cells (sperm in males and ova in females) become available so that the individual is capable of reproduction. Other physical characteristics, such as facial hair in males and breasts in females, also appear, and a growth spurt takes place so that adult height is achieved. Meanwhile, vital personal development takes place. As childhood ends, the individual becomes more self-aware and sophisticated, and independence and an adult identity are forged through a process that can be emotionally turbulent. By the end of the late teens, the individual has progressed both physically and personally, and is ready for adult life.

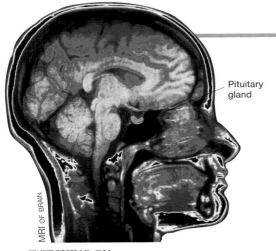

MRI OF BRAIN

SWITCHING ON
Puberty is triggered by the hypothalamus (see p. 95) which sends hormones to the pituitary gland, located at the base of the brain. The pituitary gland then releases hormones of its own that stimulate the ovaries and testes to mature. As these organs become active, they too secrete hormones. These cause changes such as female breast growth and the deepening male voice.

Pituitary gland

PUBERTY FOR BOYS
In boys, puberty usually begins between the ages of 12 and 14, although this varies greatly. The first sign of puberty is enlargement of the testes as they prepare to produce sperm. The penis then begins to grow, and reaches its full adult size after about two years. Pubic, facial, armpit, and chest hair grows. Structures of the larynx (voice box) increase in size, making the voice deeper. As sexual development progresses, the growth rate accelerates, the shoulders broaden, and muscles develop.

FACIAL HAIR
An increase in the hormone testosterone causes facial hair to grow as boys progress through puberty. This photo (left) shows facial hairs that have been shaved and are now growing back. The first hairs are fluffy, but later become thicker and stronger.

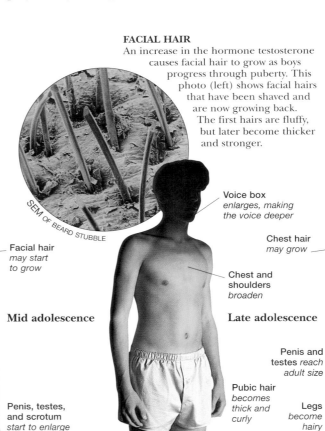

SEM OF BEARD STUBBLE

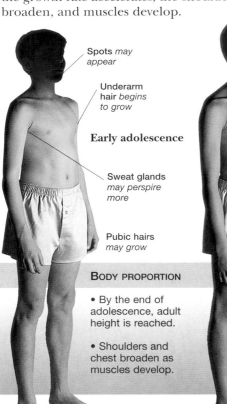

Spots *may appear*

Underarm hair *begins to grow*

Early adolescence

Sweat glands *may perspire more*

Pubic hairs *may grow*

Facial hair *may start to grow*

Mid adolescence

Penis, testes, and scrotum *start to enlarge*

Voice box *enlarges, making the voice deeper*

Chest and shoulders *broaden*

Chest hair *may grow*

Late adolescence

Penis and testes *reach adult size*

Pubic hair *becomes thick and curly*

Legs *become hairy*

BODY PROPORTION

• By the end of adolescence, adult height is reached.

• Shoulders and chest broaden as muscles develop.

SEXUAL ORGANS

• Penis, testes, and scrotum enlarge gradually.

• The testes begin to produce sperm.

• Scrotum gets darker or redder in colour.

OTHER CHANGES

• Hair grows on face, chest, pubic area, and legs.

• Voice deepens.

• Perspiration glands become more active, and skin may develop acne.

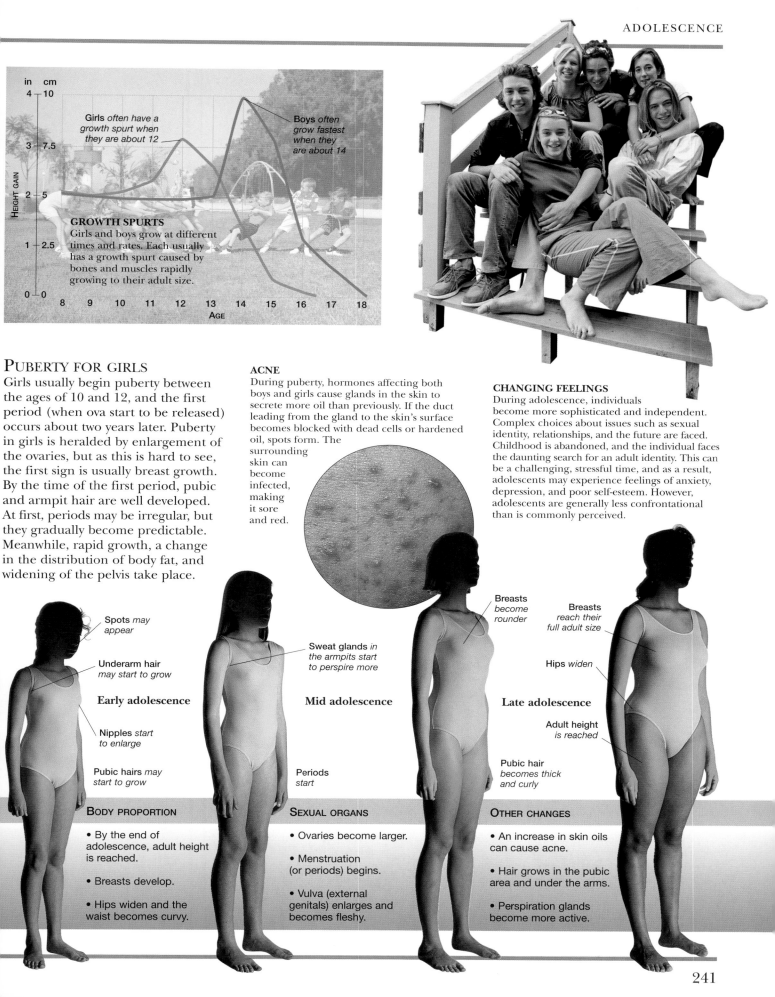

GROWTH SPURTS

Girls *often have a growth spurt when they are about 12*

Boys *often grow fastest when they are about 14*

GROWTH SPURTS
Girls and boys grow at different times and rates. Each usually has a growth spurt caused by bones and muscles rapidly growing to their adult size.

PUBERTY FOR GIRLS

Girls usually begin puberty between the ages of 10 and 12, and the first period (when ova start to be released) occurs about two years later. Puberty in girls is heralded by enlargement of the ovaries, but as this is hard to see, the first sign is usually breast growth. By the time of the first period, pubic and armpit hair are well developed. At first, periods may be irregular, but they gradually become predictable. Meanwhile, rapid growth, a change in the distribution of body fat, and widening of the pelvis take place.

ACNE

During puberty, hormones affecting both boys and girls cause glands in the skin to secrete more oil than previously. If the duct leading from the gland to the skin's surface becomes blocked with dead cells or hardened oil, spots form. The surrounding skin can become infected, making it sore and red.

CHANGING FEELINGS

During adolescence, individuals become more sophisticated and independent. Complex choices about issues such as sexual identity, relationships, and the future are faced. Childhood is abandoned, and the individual faces the daunting search for an adult identity. This can be a challenging, stressful time, and as a result, adolescents may experience feelings of anxiety, depression, and poor self-esteem. However, adolescents are generally less confrontational than is commonly perceived.

Spots *may appear*

Underarm hair *may start to grow*

Nipples *start to enlarge*

Pubic hairs *may start to grow*

Early adolescence

Sweat glands *in the armpits start to perspire more*

Periods *start*

Mid adolescence

Breasts *become rounder*

Late adolescence

Breasts *reach their full adult size*

Hips *widen*

Pubic hair *becomes thick and curly*

Adult height *is reached*

BODY PROPORTION

• By the end of adolescence, adult height is reached.

• Breasts develop.

• Hips widen and the waist becomes curvy.

SEXUAL ORGANS

• Ovaries become larger.

• Menstruation (or periods) begins.

• Vulva (external genitals) enlarges and becomes fleshy.

OTHER CHANGES

• An increase in skin oils can cause acne.

• Hair grows in the pubic area and under the arms.

• Perspiration glands become more active.

241

Adulthood and ageing

An individual's personality and lifestyle evolve as he or she progresses through adulthood. The way in which this takes place is unique to everyone, and is influenced by factors such as personal preference, opportunities and circumstances, and social and cultural expectations. Indeed, the lifestyles of two adults of similar age often differ just as much as those of two people from separate generations. As an individual's life progresses, physical changes also take place. The dramatic development of childhood and adolescence is replaced by a subtler, more gradual transition, and in due course signs of ageing begin to appear in the bones, skin, senses, and other organs. Despite these changes, modern medicine allows many elderly people to have an enjoyable and healthy old age.

COLOURED X-RAY OF A LOWER SPINE WITH OSTEOPOROSIS

Osteoporosis *has caused this vertebra to collapse*

ADULTHOOD

The nature of adulthood varies enormously depending on personality, decisions made, circumstances, and social context. However, commonly this is a time of independence and responsibility. Adulthood allows personal freedom but this is often accompanied by duties – adults are generally held accountable for their actions, and are expected to support themselves and their families financially. Meanwhile, choices must be made about careers, relationships, and parenthood, all of which involve taking on responsibilities.

MENOPAUSE

The menopause is when a woman stops having periods, usually between the ages of about 45 and 54. The menopause represents the end of a woman's reproductive potential, and this can have emotional and psychological impacts. Also, as the ovaries cease to work, levels of the hormone oestrogen decline. This may cause symptoms such as hot flushes, loss of sexual desire, and wrinkling of the skin, and also increases the risk of disorders such as stroke and osteoporosis. Women may therefore choose to use hormone replacement therapy, in which artificial oestrogen is taken in pills (right) or through patches (top right).

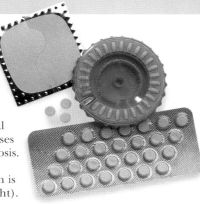

BONE DEGENERATION
This X-ray shows the spine of a 65-year-old woman who has developed a bone condition called osteoporosis. This disease causes the outer layer of bones to become thinner, while the inner layer becomes more porous, with fewer blood cells and less calcium. Bones are therefore more brittle and likely to be damaged. Here, a vertebra has become compressed and wedge-shaped, distorting the spine. Osteoporosis is common in older women, particularly after the menopause.

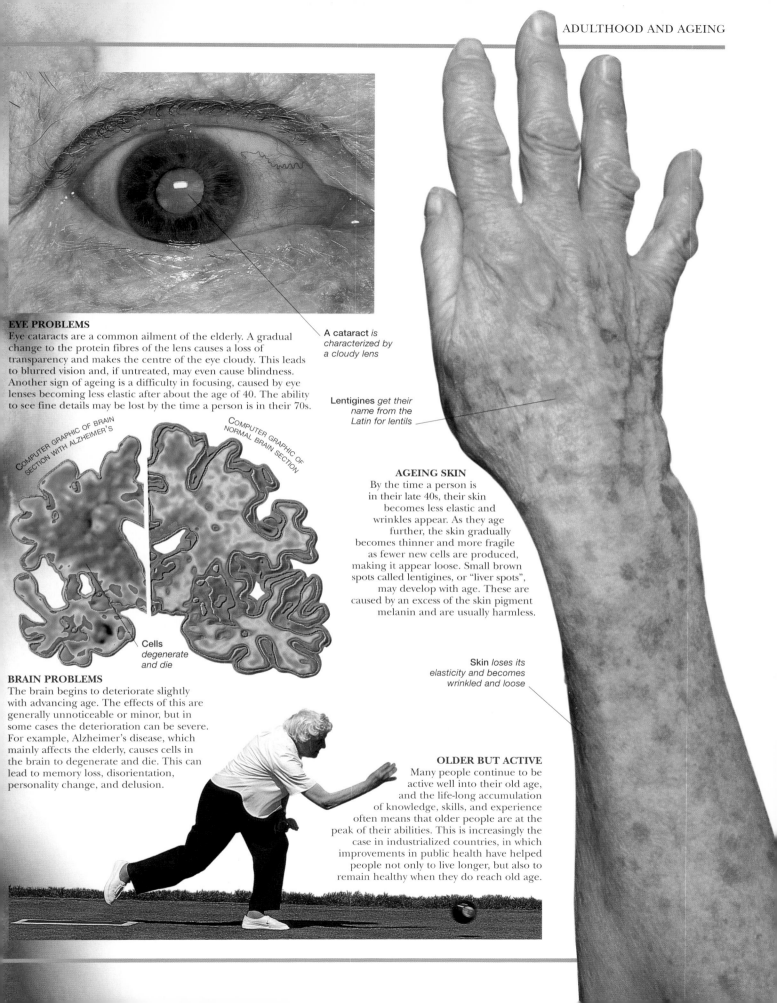

A cataract *is characterized by a cloudy lens*

EYE PROBLEMS

Eye cataracts are a common ailment of the elderly. A gradual change to the protein fibres of the lens causes a loss of transparency and makes the centre of the eye cloudy. This leads to blurred vision and, if untreated, may even cause blindness. Another sign of ageing is a difficulty in focusing, caused by eye lenses becoming less elastic after about the age of 40. The ability to see fine details may be lost by the time a person is in their 70s.

COMPUTER GRAPHIC OF BRAIN SECTION WITH ALZHEIMER'S

COMPUTER GRAPHIC OF NORMAL BRAIN SECTION

Lentigines *get their name from the Latin for lentils*

AGEING SKIN

By the time a person is in their late 40s, their skin becomes less elastic and wrinkles appear. As they age further, the skin gradually becomes thinner and more fragile as fewer new cells are produced, making it appear loose. Small brown spots called lentigines, or "liver spots", may develop with age. These are caused by an excess of the skin pigment melanin and are usually harmless.

Cells *degenerate and die*

BRAIN PROBLEMS

The brain begins to deteriorate slightly with advancing age. The effects of this are generally unnoticeable or minor, but in some cases the deterioration can be severe. For example, Alzheimer's disease, which mainly affects the elderly, causes cells in the brain to degenerate and die. This can lead to memory loss, disorientation, personality change, and delusion.

Skin *loses its elasticity and becomes wrinkled and loose*

OLDER BUT ACTIVE

Many people continue to be active well into their old age, and the life-long accumulation of knowledge, skills, and experience often means that older people are at the peak of their abilities. This is increasingly the case in industrialized countries, in which improvements in public health have helped people not only to live longer, but also to remain healthy when they do reach old age.

HUMAN INHERITANCE

IT IS OFTEN EASY TO recognize members of a family because of the remarkable similarity between them. Just as striking, though, are the differences. Why individuals inherit some features but not others has been studied over the past century by scientists. The key to inheritance lies inside body cells, where chromosomes store the information needed to construct and run each cell in "units" called genes. Since cells make up the body, they also determine its appearance. When humans reproduce, the chromosomes in their sperm and ova are passed on to the next generation.

FAMILY TREE
This photo shows four generations from the same family. Each individual developed from a zygote, or fertilized ovum, produced when their father's sperm fertilized their mother's ovum. Both ovum and sperm have the same number of chromosomes, but some of the genes (genetic instructions) that they carry differ. Passing on a new combination of genes to an offspring means that, while one generation resembles the previous one, a daughter is not identical to her mother (or father). Everyone, therefore, apart from identical twins, has a unique combination of genes in their chromosomes.

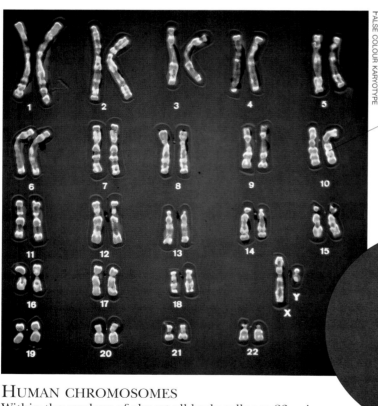

FALSE COLOUR KARYOTYPE

Each chromosome pair *contains many genes, with instructions for particular characteristics*

X chromosomes *of a female match each other*

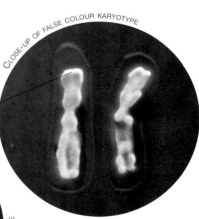

CLOSE-UP OF FALSE COLOUR KARYOTYPE

Female sex chromosome

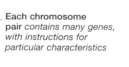

CLOSE-UP OF FALSE COLOUR KARYOTYPE

Male sex chromosome

X and Y chromosomes *of a male are of different length*

SEX CHROMOSOMES
This single pair of chromosomes is responsible for determining a person's sex. Not surprisingly, they differ between females and males. A female's body cells each contain a matching pair of sex chromosomes called X chromosomes. A male's sex chromosomes do not match. His body cells each contain one X chromosome paired with a Y chromosome.

HUMAN CHROMOSOMES
Within the nucleus of almost all body cells are 23 pairs of thread-like chromosomes. Normally, chromosomes are difficult to see under the microscope because they are long, thin, and tangled. But they can be seen when a cell is about to divide because they get shorter and thicker. A photograph of the chromosomes is cut up to produce a karyotype like this one. Pairs of matching chromosomes are arranged in order of size from 1 (largest) to 22 (smallest). These 22 pairs of chromosomes, called autosomes, control most body characteristics. The remaining pair are the sex chromosomes.

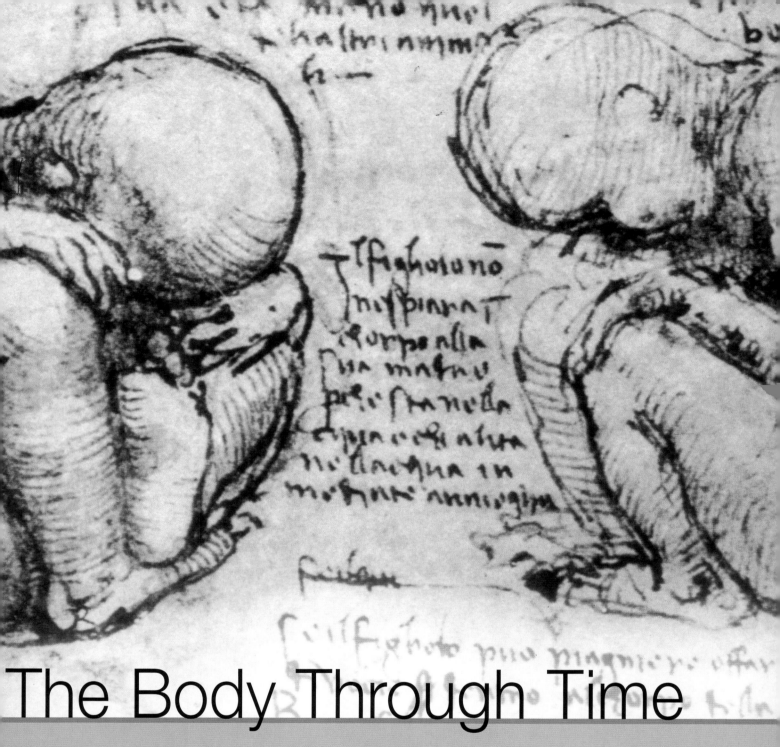

The Body Through Time

THEIR POWERFUL INTELLIGENCE, and ability to analyse, communicate, and record, make human beings unique in the living world. Driven by a natural curiosity, these skills have allowed humans to study themselves in order to understand both how the body works and why it goes wrong. Over the millennia, human biology and medicine have evolved together across many cultures. The resulting accumulation of knowledge enables 21st-century humans to stay healthier and live longer than their ancient relatives.

Early humans

SCIENTIFIC MEASUREMENTS OF SKULLS show that larger brain size evolved quickly in ancient human relatives, while walking on two legs instead of four left their hands free to use stone tools. Increasing intelligence and skill in tool use made early humans formidable hunter-gatherers. They may even have learned by trial and error to use herbs to treat illnesses, in the same way as some apes are thought to do today. Combined use of intelligence and hands also allowed them to produce early works of art, revealing an awareness of themselves and their surroundings, which is a characteristic that distinguishes humans from apes.

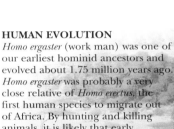
Proconsul
20 million years ago

ANCIENT RELATIVES
This chart shows the evolutionary sequence from an ape-like ancestor, *Proconsul*, that walked on all fours, to ancient hominid (human-like) relatives and modern humans. The sequence is not a straight line, but a "tree" with dead-ends that do not necessarily indicate ancestry.

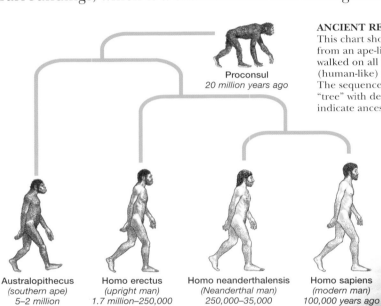

Australopithecus
(southern ape)
5–2 million
years ago

Homo erectus
(upright man)
1.7 million–250,000
years ago

Homo neanderthalensis
(Neanderthal man)
250,000–35,000
years ago

Homo sapiens
(modern man)
100,000 years ago

HUMAN EVOLUTION
Homo ergaster (work man) was one of our earliest hominid ancestors and evolved about 1.75 million years ago. *Homo ergaster* was probably a very close relative of *Homo erectus*, the first human species to migrate out of Africa. By hunting and killing animals, it is likely that early humans would have learned about the heart and other organs.

TRACES OF OUR ANCESTORS

Humans belong to the genus *Homo*, and evolved from their ape-like ancestor, *Proconsul*. All early humans would have been hunter-gatherers, living on a diet of plants and animals. Neanderthals, an ancient relative of modern *Homo sapiens*, appeared about 250,000 years ago. Archaeological evidence supports the argument that they looked after their sick and probably knew how to set broken bones: the remains of a Neanderthal man found in a cave in Shanidar, Iraq, show crushing injuries and yet he lived to tell the tale. This is deduced from the fact that the bones show signs of healing. Neanderthals also buried their dead. In one Shanidar burial, pollen from seven different flowers was found. Six of these plants have medicinal properties. Was this a coincidence or were these ancient relatives using medicinal plants?

30,000
years ago

40,000
years ago

100,000
years ago

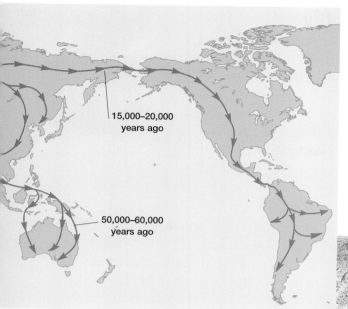

15,000–20,000
years ago

50,000–60,000
years ago

OUT OF AFRICA

The distribution of genes in modern humans shows that *Homo sapiens* evolved in East Africa. They migrated out of Africa about 100,000 years ago. They then travelled across Asia, Europe, Australia, and into North America, eventually reaching South America.

Area of
healed bone

EVIL SPIRITS

This preserved skull shows three trepanned holes for allowing evil spirits to escape – a previous hole has healed, which suggests that the person survived the practice. Cave paintings showing men dressed as animals may represent shamans driving away the demons.

MODERN HUMANS

Our ancient ancestors would have used sounds and facial expressions to communicate desires and emotions, just as humans do today. Although they would have had some understanding of the use of plants, they did not know how the body worked and would have explained illnesses in terms of evil spirits. Cave paintings left by *Homo sapiens* are difficult to interpret. But it is likely that the cave painting above right shows a shaman (healing priest) dancing or chanting to drive away evil and illness – shamans are still found today, practising their blend of priest and doctor duties. There is also evidence to suggest that by 5000 BC, *Homo sapiens* carried out trepanning operations: holes were made in the skull to release the evil spirits believed to have caused the illness.

NATURAL REMEDIES

By watching apes, we can find clues to the ways in which early medicine might have begun. Chimpanzees deliberately pick and chew leaves of certain plants, then spit them out. Have they learned that some plants can cure sickness? This might be the way in which our human ancestors learned to use plants as medicines. Unlike apes, humans developed sophisticated language skills, allowing them to pass on this knowledge to their young.

INSTINCTS

Gorillas in West Africa eat several types of plant that are known to help joint pains and other conditions.

The ancient world

MODERN MEDICAL TECHNIQUES have been used to analyze mummified bodies from ancient tombs. The results show that the people of ancient civilizations suffered from illnesses that are still prevalent today, such as gallstones and appendicitis. In the ancient world, sick people believed that they were being punished by their gods for wrong-doing. Although they turned to physicians for treatments, they often put their faith in the hands of priests, who used magic and called upon the gods to take pity and deliver a cure.

IMHOTEP
One of the first great physicians in Egypt was Imhotep (c.2686–2613 BC), who was elevated to the status of a god.

MESOPOTAMIAN MEDICINE

Physicians practised in Mesopotamia – the land between the Rivers Tigris and Euphrates, and home to the Sumerian, Assyrian, and Babylonian civilizations. A Sumerian stone tablet dating from 2150 BC shows how physicians provided first aid by washing and binding wounds, and applied poultices made from prunes, pine, and lizard dung. In Assyria, priests were also physicians. They learned their skills from thousands of inscribed stone tablets, such as the one above. Professional physicians existed by the reign of King Hammurabi of Babylon, who ruled from 1792 to 1750 BC. He established a set of laws called the Code, which stated that surgeons were responsible for their errors. There were rewards for surgeons who cured noblemen and severe punishments, such as cutting off hands, for surgeons whose operations killed patients of high status. But if a slave died, the surgeons only had to replace the slave. Religion also played an important part in Babylonian medicine. This took the form of ritual incantations and appealing to the heavens for help in curing diseases.

ASSYRIAN STONE TABLET
This stone tablet from the library of Ashurbanipal (668–627 BC), the last great Assyrian ruler, contains advice on remedies for stomach disorders.

EGYPTIAN MEDICINE

Physicians in ancient Egypt were undoubtedly the best doctors of the day. Even kings of Babylonia and of the Hittites wrote to the pharaohs, asking for Egyptian physicians to be sent to treat them. The Ebers papyrus, the oldest medical text, contains remedies that included hippopotamus dung and onions. But such ingredients were only used when all else failed. Surgeons, who were also priests, carried out a limited amount of internal surgery, while providing spiritual comfort to the sufferer.

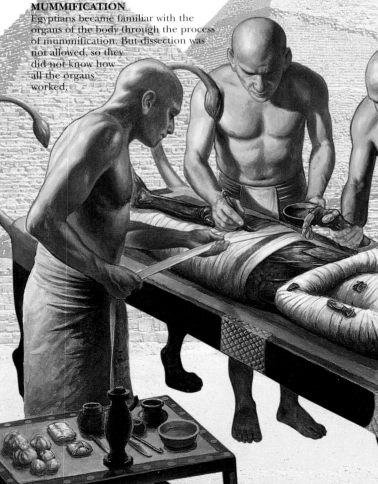

MUMMIFICATION
Egyptians became familiar with the organs of the body through the process of mummification. But dissection was not allowed, so they did not know how all the organs worked.

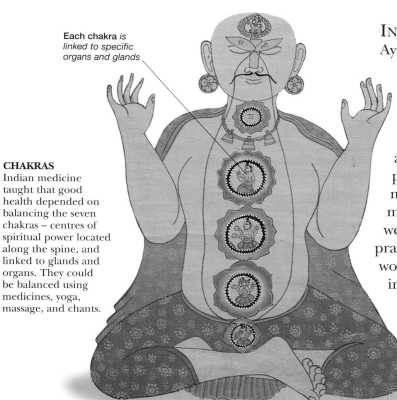

Each chakra *is linked to specific organs and glands*

CHAKRAS
Indian medicine taught that good health depended on balancing the seven chakras – centres of spiritual power located along the spine, and linked to glands and organs. They could be balanced using medicines, yoga, massage, and chants.

INDIAN MEDICINE

Ayurveda, which means "knowledge of life", is a system of medicine that was used in ancient India and is still in use today. It treats illnesses by correcting imbalances in the body's three humours – wind (vaya), bile (pitta), and phlegm (kapha). It is likely that the ancient Greek theory of the four humours (see pp. 258–9) orginated in India with Ayervedic medicine. Physicians used chants together with medicine and operations. Surgery was particularly well developed in ancient India, where practitioners used fine steel needles to sew up wounds, with biting black ants as sutures to close internal wounds. Susruta, who lived some time between 100 BC and AD 100, carried out cosmetic surgery, repairing amputated noses by using skin grafted from the forehead. Trainee surgeons often practised their skills on meat and leather bags filled with slime before working on real patients!

CHINESE MEDICINE

Harmony and balance were also key elements of Chinese medicine, which by 400 BC had become separated from religion. According to Chinese medical philosophy, harmony between two opposing forces, yin and yang, is essential for good health. The great reference book of Chinese medicine, the *Nei Ching*, was compiled between 479 and 300 BC, based on the wisdom of Emperor Huang Ti (2629–2598 BC). Many of the techniques it describes have survived in traditional medicine in China to the present day. It includes details of using acupuncture to balance yin and yang by controlling the flow of energy (chi) in the body.

YIN AND YANG
The two fundamental forces in Chinese philosophy – including medicine – are yin and yang. Yin is dark, cool, moist, passive, negative, and female, while yang is light, dry, active, warm, positive, and male. It is believed that illness is caused by an imbalance of these forces.

PULSE TAKING
Taking the pulse gave Chinese physicians information on how well the chi energy of the body was balanced. Six pulses were counted on each wrist, and described in poetic phrases before treatment was prescribed.

Ancient Greece and Rome

THE GREEKS WERE INFLUENCED CONSIDERABLY by the Egyptians, and even identified their god of medicine, Asklepios, with the Egyptian god of medicine, Imhotep. Greek medicine was based on a combination of supernatural and natural beliefs. The Greeks prayed to Asklepios for a cure and went to the temples to be healed. However, Hippocrates believed that a disease was not sent or cured by the gods, but had a natural cause and cure. Roman medicine had many similarities, but with emphasis on the importance of preventing disease by providing clean water and sewers in crowded cities. The Romans also adopted the ideas of Greek-born Claudius Galen.

CLAUDIUS GALEN
The influence of the writings of Claudius Galen (c.AD 130–200) lasted until the Renaissance. Many of his teachings contained mistaken ideas about human anatomy.

THE BASIS OF GREEK MEDICINE

Greek medicine taught that illnesses were caused by a lack of balance between four body fluids called humours. These were blood, phlegm, yellow bile, and black bile. The physician Hippocrates (c.460–377 BC) founded a school of medicine on the island of Cos, where pupils learned to diagnose and treat diseases by rebalancing humours. His Hippocratic Oath, which committed doctors to maintaining confidentiality in their consultations with patients and to acting only in the patient's best interests, is still followed by doctors today.

THE FOUR HUMOURS
The Greeks developed the idea that health was controlled by the balance of four bodily humours. Excess blood was removed by bleeding, while other humours were altered by prescribing herbs or a change in diet.

GREEK PHYSICIANS

Patients visiting a Greek physician would be closely questioned about their symptoms, diet, and lifestyle, and were provided with a prognosis, advising them on the likely progress of their illness. Dissection of human bodies was not allowed in Greece, but was practised in the city of Alexandria, founded by the Greeks in Egypt. There Herophilus (c.300 BC), a skilled anatomist, learned to identify arteries and nerves, and investigated the structure of the brain.

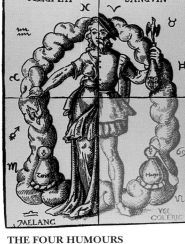

GREEK DOCTORS
Greek doctors followed the teachings of Hippocrates, who separated magic from medicine. They diagnosed symptoms, using methods that included feeling for palpitations and listening to sounds from the patient's chest.

ROME'S DEBT TO GREEK MEDICINE

When Rome was founded, its people placed great reliance on gods for healing, and when plague struck in 293 BC they called in help from the Greek god Asklepios. They adopted him as their own god, and welcomed Greek physicians. Doctors enjoyed privileged status in Rome, free from military service and taxes, so there were many practitioners. There were also women doctors, who worked as midwives and specialists in female disorders.

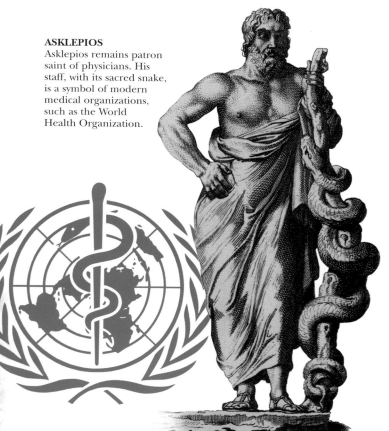

ASKLEPIOS
Asklepios remains patron saint of physicians. His staff, with its sacred snake, is a symbol of modern medical organizations, such as the World Health Organization.

ROMAN SURGERY
The army of Julius Caesar (c.100–44 BC) included battlefield surgeons who set up dressing stations. Injured soldiers were sent back to field hospitals *(valetudinaria)*, the forerunners of today's hospitals. Roman surgeons were skilled in operations ranging from removal of splinters to amputation of limbs.

DEVELOPMENT OF ROMAN MEDICINE

The Roman army was responsible for many of the important developments in medicine. Military campaigns had taught Roman doctors about the nature of wounds and surgery, and they also used gladiators' wounds as "windows into the body". Cornelius Celsus (c.20 BC–AD 45) compiled an encyclopedia, *De Medicina*, that described in Latin everything there was to know about medicine. It provided advice on clamping veins, so casualties would not bleed to death, and included details of delicate operations, such as eye surgery and the removal of bone splinters from head wounds.

GALEN AND HIS LEGACY

Claudius Galen was the most influential physician in Rome. But, prevented from dissecting humans, he studied anatomy by dissecting pigs and other animals. Although he made some significant discoveries, such as nerve function, many of his ideas were wrong, such as the reliance on blood-letting. But because they were backed by the Christian church, Galen's ideas would dominate medicine for the next 1,500 years.

Arabic medicine

WHEN BARBARIANS DESTROYED THE ROMAN EMPIRE and medicine fell into decline, sick people in Western Europe once again placed their faith in religion and magic. The teachings of Galen and the ancient Greeks survived only because they had been carried eastwards to Persia, where Arab scholars translated them into their own languages. As the new religion of Islam spread explosively through the Middle East and North Africa and into Europe, medicine flourished, based on the Greek-Roman model. Hospitals and medical schools were established in major cities, where new treatments and more advanced methods of surgery were available. In the 12th century, ancient medical knowledge, saved and enhanced by the Arabs, travelled back to Europe through Spain and Italy.

RHAZES
Although Rhazes (above) was a follower of Galen, he recommended that his teaching should not be followed slavishly, and advocated the use of reason and science in medicine.

MEDICAL ADVANCES

Physicians added new knowledge to the teachings of the Greeks and Romans. Rhazes (c.850–932) was the first to describe measles and smallpox as separate diseases, and reasoned that fever was part of the body's defence against illness. Avicenna of Persia (980–1037), also known as Ibn Sina, produced his *Canon of Medicine*, which superseded Galen's work and remained influential in medical teaching until the 17th century. Some Arabic ideas were very advanced, such as those of the Syrian physician Ibn an-Nafis (died 1288 AD), who argued that blood circulated around the body – a theory that would not be proven until the work of William Harvey, 400 years later.

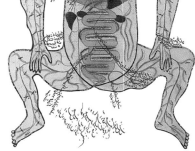

ANATOMICAL DRAWINGS
Anatomical drawings followed the teachings of Galen, and perpetuated his errors. A more accurate understanding of the human body was not possible as dissection was forbidden under Islamic law.

REMEDIES
Medicinal preparations were dispensed by physicians as well as pharmacists. Physicians trained at a *madrasah* (place of lessons), which was attached to a mosque or hospital.

Scalpel

Scissors

Camphor

Pestle

Surgical knife

Mortar

TOOLS OF THE TRADE
Surgeons used finely crafted tools for carrying out the numerous operations, while a pestle and mortar was often used for crushing substances. These would then be incorporated into other preparations, such as ointments, pills, and confections.

HERBAL TRADITIONS

During the 9th century, the first private pharmacies opened in Baghdad. Pharmacy came to be viewed as a separate profession from medicine, which was practised by skilled specialists who had to pass exams and be licensed. The use of herbs played a major role in this development. Arab pharmacists helped to introduce a number of new drugs, such as camphor, senna, and nutmeg. They were the first to develop sweet-tasting syrups for medicine, and used pleasant flavouring extracts, such as rose water and orange blossom water, as a means of administering the medicine. Eventually, extracting and preparing the medicine became a fine art.

Cautery irons *were heated like branding irons, and then placed on the affected area to seal the wound*

ARAB SURGEONS
Although surgery did not progress much, surgeons carried out complex operations, and put much faith in cauterizing. They learned their trade from illustrated manuals of surgery, such as *Al-tasrif* compiled by Albucasis of Córdoba.

SURGERY

There were three types of surgery carried out by Arab surgeons: vascular, general, and orthopaedic (ophthalmic surgery was seen as a separate speciality). Albucasis (936–1013) was the most famous surgeon in Arabic medicine. His book includes descriptions of various operations, such as correction of spinal deformities, removal of eye cataracts, and even tracheotomy. Today, surgeons still carry out several procedures introduced by Albucasis.

The Middle Ages

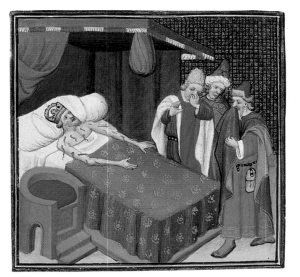

MEDIEVAL TREATMENTS
Bleeding patients to balance their humours and remove poisons remained as important in medieval medicine as it had been in Roman times. Blood was usually drained by applying leeches or cutting open veins from the part of the body closest to the source of illness.

IN THE MIDDLE AGES, between the 5th and 15th centuries, the sick throughout Western Europe accepted illnesses as God's punishment for their sins and believed that terrible plagues, such as the bubonic plague, were expressions of divine wrath. People placed more reliance on the prayers of priests than help from physicians, whose treatments often degenerated into blends of superstition and magic. Caring for the sick became an act of Christian charity, with little attempt to use medical science to find cures. Eventually, medical science was revived through the establishment of medical schools based on Arab practices and the rediscovery of Greek teachings, which had been preserved in the Arab world.

MYTH, MAGIC, BUT VERY LITTLE MEDICINE

Treatments remained much as they had been in Galen's day, with a reliance on bloodletting to balance the humours. Uroscopy, the examination of urine samples to look for disease symptoms, was an important form of diagnosis, and many treatments used herbal medicine. Physicians seeking guidance for the use of herbs looked for divine signs by following the Doctrine of Signatures – plants that resembled bodily parts were believed to have been shaped by God to draw attention to their beneficial uses. Purveyors of folk remedies also flourished, offering such unlikely cures as plucked crabs' eyes for sight defects. Major surgery was rare, but minor operations, such as cauterizing wounds with hot irons, were carried out by barbers as a lucrative side-line to shaving and cutting hair.

CARING FOR THE SICK

The Christian Church played an important part in caring for the sick during the Middle Ages. Throughout Europe, hospitals were established where the sick were fed, sheltered, and prayed for, but there were few medical advances that could offer any hope of improved cures. Monks often acted as physicians and followed those parts of Galen's teaching that fitted comfortably with their religious beliefs.

CHRISTIAN CHARITY
Hospitals offered their patients care and comfort, but beyond the reliance on prayers, there was no method of treating disease.

TREATING PATIENTS

Although treatments for illnesses did not keep pace with advances in understanding how the body works, they did become more scientific. The English physician Thomas Sydenham (1624–89) was one of the most respected physicians of the age, noted for his descriptions of diseases and prescription of specific medicines for treating them. He held the view that physicians should put their patients before testing theories, and relied on common sense and observation. Understanding of herbal medicines improved during this period. Sydenham was the first to prescribe Peruvian bark (the source of quinine) to treat malaria. Dutch physician Herman Boerhaave (1668–1738), who greatly admired Sydenham's ideas, established a reputation for encouraging careful scientific diagnosis and recording detailed case histories. Some of his ideas are still used in today's teaching methods, such as the involvement of students in post-mortems as a way of learning about the cause of death.

FIRST MICROSCOPES
The invention of microscopes allowed scientists to explore beyond the limits of the human eye, showing that organs were made of tissues and cells. This microscope was made for Robert Hooke in the late 1600s.

Eyepiece

Water flask

Specimen

Oil lamp

MEDICINE AS A PROFESSION

The status of surgeons rose as knowledge of the workings and defects of the internal organs increased, but surgery remained a frightening and risky ordeal. Lack of anaesthetics meant that conscious patients screamed in agony when surgeons operated, and abdominal surgery was rare because many patients died of post-operative infections. But, by the end of the 17th century, the scientific approach to anatomy, surgery, and the treatment of illness was firmly established. Medicine in all its forms was becoming a respected profession.

ANATOMY LESSONS
Lessons in anatomy became an essential part of medical training in the 17th century, just as they are today. The period also saw the founding of academies of science in Rome, Paris, and London, where advances in medical science were hotly debated.

Anatomy Lesson of Dr Nicolaes Tulp **(1632) by Rembrandt van Rijn**

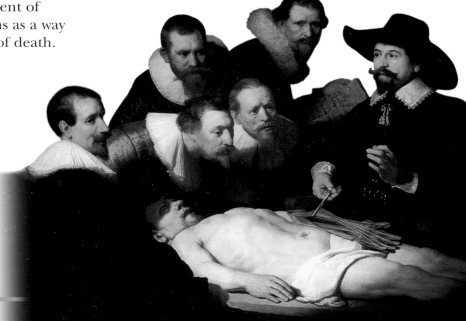

Advances in surgery

JOHN HUNTER
John Hunter turned surgery into a science. He dedicated his career to comparative anatomy and experimental surgery.

AT THE BEGINNING OF THE 19TH CENTURY, people who discovered that they needed an amputation received the news almost as a death sentence. Enduring the pain, at a time when no effective anaesthetics were available, was bad enough. The best that they could hope for was a surgeon who worked swiftly. And after the operation, wounds always became septic, so as many as half of all amputees died. By the end of the century, two major advances – the introduction of anaesthetics and the use of antiseptics – had drastically reduced the trauma of surgery and the risk of death from infections.

SWIFT, AGONIZING OPERATIONS
During the 18th century, brothers John (1728–93) and William Hunter (1718–83) transformed surgery from a lowly trade into a medical science. But for patients it remained a subject of dread and despair. They remained fully conscious throughout major operations, often screaming in agony. Many surgeons prided themselves in being able to work quickly, cutting off a leg in less than three minutes and so minimizing their patients' suffering.

PAINLESS SURGERY
Several dentists experimented in the 1840s with ether as an anaesthetic for numbing pain during tooth extractions. In 1846, American dentist William Morton (1819–68) persuaded a surgeon called John Collins Warren (1778–1856) to use ether on a patient during the removal of a neck tumour at Massachusetts General Hospital, USA. The unconscious patient barely murmured during the 25-minute operation. The era of painless surgery had arrived. The introduction of anaesthetics allowed surgeons to operate slowly and carefully, and to attempt complicated operations that were impossible when patients struggled in agony.

SURGICAL INSTRUMENTS
It is no coincidence that these instruments (right), used for amputations, look more like a carpenter's tools. Until the 18th century, surgery was a craft, not a science. Although surgical instruments were made of polished wood and steel to allow easy cleaning, surgeons did not sterilize instruments between operations.

FAST WORK
Before anaesthetics became available, patients cried out in pain throughout the operation, and had to be tied down, as shown above. The need to minimize patients' suffering meant that surgeons achieved fame for speed, not precision.

KILLER INFECTIONS

Before the mid-19th century, deaths from septic infections after surgery were so common that patients had less chance of survival than a soldier at the Battle of Waterloo. Hungarian doctor Ignaz Semmelweiss (1818–65) showed that deaths amongst new mothers could be reduced if medical staff washed their hands, but surgeons rarely did this, or sterilized instruments (see pp. 232–3). The turning point came in 1865, when Joseph Lister (1827–1912) introduced carbolic acid spray to kill germs. Until Lister's day, surgeons did not realize that bacteria caused infections. Like Galen (see p. 259), they believed that pus in septic wounds was part of the natural healing process.

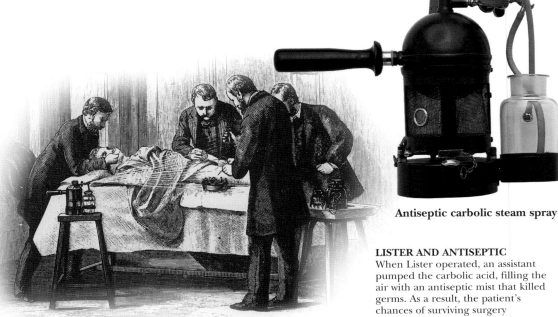

ANAESTHESIA
William Morton pursuades John Warren to use ether as an anaesthetic during surgery, stopping the patient from feeling pain.

THE THEATRE

Operating theatres were built as theatres, where students – sometimes coughing and sneezing – sat in tiers of seats, watching surgeons perform. Lister insisted on thorough cleaning of operating theatres, so he used carbolic acid spray

to kill germs. His first great success came when he saved the leg of an 11-year-old boy, who had been run over by a cart. The compound fracture of the boy's leg was so bad that the leg would normally have been amputated immediately, but Lister set the fracture and sterilized it with carbolic acid. The boy's leg healed, enabling him to walk out of the hospital within six weeks. Lister struggled to overcome doubts from rival surgeons, but when he successfully operated on Queen Victoria using carbolic acid spray, his reputation was made, and antiseptics became a feature of surgical operations.

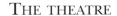

Antiseptic carbolic steam spray

LISTER AND ANTISEPTIC
When Lister operated, an assistant pumped the carbolic acid, filling the air with an antiseptic mist that killed germs. As a result, the patient's chances of surviving surgery dramatically increased.

Microbe hunters

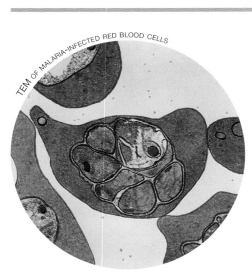

TEM OF MALARIA-INFECTED RED BLOOD CELLS

MALARIAL MICROBE
These red blood cells are infected with *Plasmodium*, the single-celled parasite identified in 1880 by Charles Laveran as the cause of malaria. Infected cells rupture periodically, releasing parasites that invade other red blood cells, and causing the high fever that is typical of the disease.

THROUGHOUT RECORDED HISTORY, humankind has been plagued by diseases. But, at the beginning of the 19th century, doctors and scientists still had no clear idea of what caused diseases apart from rather vague notions of poor hygiene or bad air. One hundred years later, everything had changed, thanks to two pioneering scientists: French bacteriologist Louis Pasteur (1822–95) and German doctor Robert Koch (1843–1910). Together, Pasteur and Koch developed the "germ theory" of disease. This states that it is microbes – micro-organisms such as bacteria – that are the cause of many diseases. Once germ theory had been established, microbe hunters rapidly identified the causes of prominent diseases, including cholera, tuberculosis (TB), and malaria. Other researchers investigated how diseases are passed from person to person.

MIASMAS AND MISUNDERSTANDING

Although Dutch microscopist Antoni van Leeuwenhoek (see pp. 16–17) first described the micro-organisms now called bacteria in 1683, this had little impact on medicine. By the mid-19th century, many doctors still held that diseases were caused by miasma, a mysterious toxic vapour that arose from stagnant water, slums, and faeces. Although doctors knew that bacteria occurred in wounds and diseased tissues, it was generally assumed that their presence was incidental and could be explained by "spontaneous generation" – the belief that new life could arise from decaying material, just as maggots seemed to "appear" on rotting meat.

LOUIS PASTEUR
The ground-breaking work of this pioneer microbe hunter simultaneously established the germ theory of disease and demolished the old notion of spontaneous generation. Although he trained to be a chemist, and had been a professor of chemistry, Pasteur was to become the first ever bacteriologist – a scientist who studies bacteria.

MIASMAS
The theory that these mysterious vapours caused disease was still widely believed in the mid-19th century.

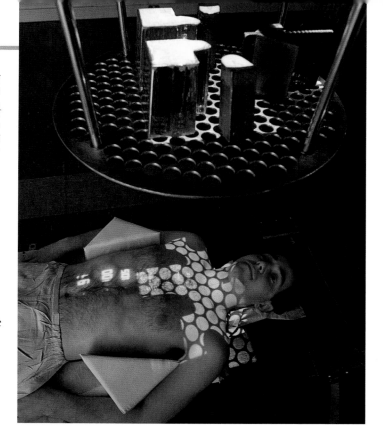

BEATING BACTERIA

Antibiotics were the most important 20th-century drug discoveries, because they kill a broad range of bacteria. Bacterial infection of burns and wounds was a major cause of death until antibiotics became available. The first, penicillin, was discovered by Alexander Fleming (1885–1955) in 1928, but only went into large-scale production in 1938, just in time to save millions of lives during the Second World War. Preventing diseases is as important as curing them, especially when caused by bacteria that can become resistant to antibiotics. Aseptic techniques in surgery, which demand absolute cleanliness, have become vital in preventing these infections and ensuring that antibiotics remain a major weapon against bacteria.

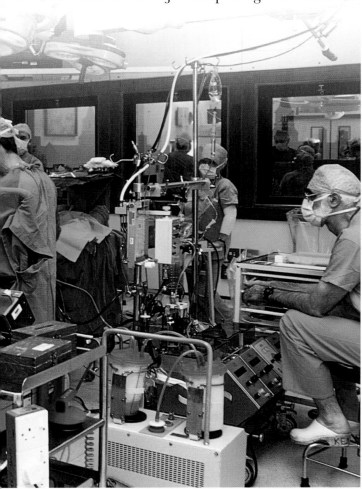

TREATMENTS AND ARTIFICIAL ORGANS

Today many more illnesses can be cured, thanks to accurate diagnosis and treatments that are more precisely targeted at the source of diseases. Techniques such as radiotherapy, using high doses of focused radiation to kill diseased cells, can be combined with drug treatments to cure diseases such as cancers, which would once have been fatal. Worn and damaged body parts can also be replaced with artificial joints, while advances in medical technology, using microelectronics and microprocessors, have produced artificial limbs that mimic arms and hands. And while organ transplant technology can replace damaged hearts, the shortage of donor organs has led to the development of the first mechanical hearts, which can prolong the life of a patient until a donor organ becomes available.

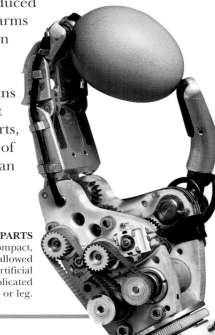

REPLACEMENT PARTS
Microelectronic circuits and compact, long-lasting batteries have allowed medical technologists to develop artificial limbs, which can perform the complicated movements of a real hand, arm, or leg.

Alternative therapies

TODAY, IF PEOPLE ARE ILL, we expect modern medical science to provide a cure, which might come in the form of a drug or drastic surgery. Our ancient ancestors had to rely on simpler treatments. These included meditation, herbal remedies, and treatments applied to the body surface in an effort to promote better health. These ancient medical methods have survived and can be used, in combination with modern alternative therapies, such as osteopathy and aromatherapy, to complement modern medicine.

Map of the body's acupuncture points

| Chamomile plant | Mother tincture | Diluted tincture | Further diluted tincture |

LIKE CURES LIKE

German doctor Samuel Hahnemann devised the therapy called homeopathy, based on the claim that if high doses of a substance produce symptoms resembling an illness, then small doses of the same substance could be used to cure that illness. Homeopathic treatments are based on plants, such as chamomile, mineral, and animal products, diluted hundreds of thousands of times.

COMPLEMENTARY MEDICINE

Modern science-based medicine has produced great benefits for humanity, eliminating diseases such as smallpox, and providing effective treatments for illnesses that were once always fatal, such as diabetes. But many people argue that the achievements of modern medicine should not eclipse the benefits of traditional treatments, which have much to offer in our search for health. There is a growing belief that modern medical science and the older teachings complement one another. Modern scientific medicine often uses technologically advanced treatments that target specific parts of the body, while complementary medicine treats the health of the whole – mind and body together. Many supporters of complementary medicine believe that we would not need drastic modern treatments so often if we recognized the benefits of traditional Chinese and Indian medicine, combined with more recently developed alternative techniques, such as osteopathy and reflexology.

ACUPUNCTURE

In the ancient Chinese practice of acupuncture, the balance of yin and yang in the body is corrected by inserting fine needles at specific places, stimulating the body's energy channels. Different acupuncture systems have mapped as many as 600 points on the body (right), where insertion of needles may have a beneficial effect.

ANCIENT PHILOSOPHIES

The ancient medical methods of Indian Ayurvedic and Chinese medicine are still practised in their countries of origin, and many people in the Western world also choose to have their ailments treated in this way. Chinese acupuncture clinics use the insertion of needles in places that influence one or more of the 12 major energy pathways in the body.

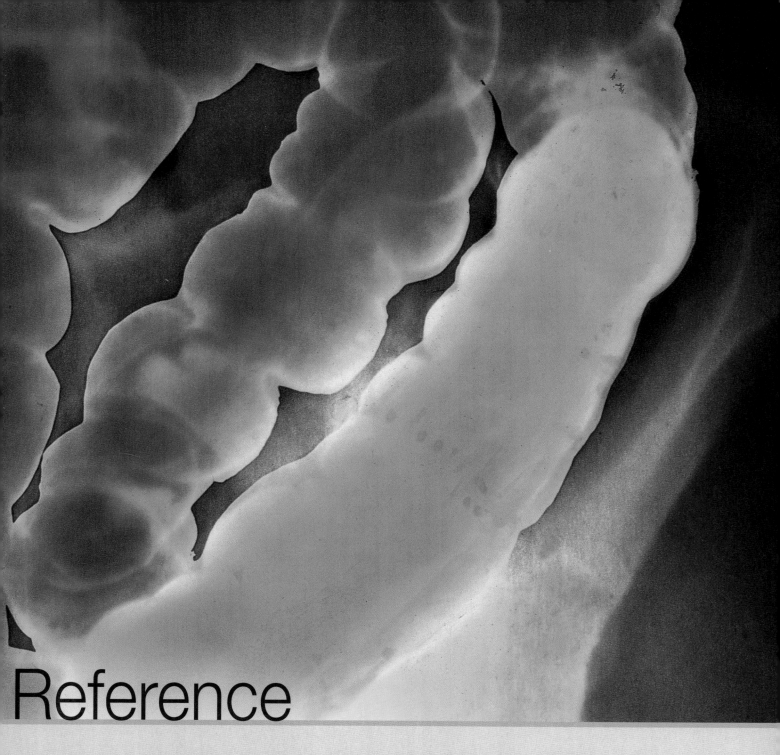

Reference

THIS FINAL SECTION PROVIDES invaluable reference resources in the form of a timeline and glossary. Spanning the millennia from 100,000 BC to the present day, the timeline lists significant milestones in both the study of the human body and the development of medicine. Any confusion over words used in the book can be solved by looking them up in the glossary, which gives easy-to-understand explanations of nearly 300 scientific terms, from "Abdomen" to "Zygote".

TIMELINE

c. 100,000 BC
Modern humans (*Homo sapiens*) first appear in Africa.

c. 70,000 BC
Humans spread from Africa to other continents.

c. 30,000 BC
Humans produce cave paintings and sculptures of themselves and other animals.

c. 10,000 BC
Transition from hunter-gatherer lifestyle to agricultural, settled communities.

c. 2650 BC
Earliest known physician, the Egyptian Imhotep, later given full status as a god.

c. 2600 BC
Chinese Emperor Huang Ti believed to have laid down the basic principles of the *Nei Ching*, a standard manual of Chinese medicine.

c. 1750 BC
King Hammurabi of Babylonia establishes a set of laws, called the Code of Hammurabi. The laws help to regulate the work of physicians.

c. 1500 BC
Date of origin of the Ebers papyrus (discovered in Egypt in 1873), which remains the oldest known medical text.

c. 500 BC
Greek physician and philosopher Alcmaeon of Croton proposes that the brain, and not the heart, is the organ of thinking and feeling.

c. 420 BC
Greek physician Hippocrates teaches the importance of observation and diagnosis over magic and myth in medicine.

c. 350 BC
Greek philosopher Aristotle states that the heart is the organ of feeling and intelligence, a belief held centuries before by the ancient Egyptians.

c. 280 BC
Herophilus of Alexandria reinstates the brain as the organ of thinking. He also describes the cerebrum and cerebellum, and discovers that nerves are channels of communication.

AD 40
Roman philosopher Cornelius Celsus publishes *On Medicine*, a medical handbook based on earlier Greek sources.

C. AD 200
Influential Greek-born Roman doctor Claudius Galen describes, often incorrectly, the workings of the body. With his ideas remaining unchallenged until the 1500s, few advances are made in the understanding of the human body.

AD 890–932
Persian physician Abu Bakr ar-Razi (Rhazes) produces many influential medical texts, and accurately describes measles and smallpox.

1000

c.1000
Publication of medical texts by Arab doctor Ibn Sina (Avicenna), which influences European and Middle-Eastern medicine for the next 500 years.

c.1000
Arab surgeon al-Zahrawi (Albucasis) publishes surgery textbooks that describe complicated operations.

1200

1268
Roger Bacon, an English scientist, records the use of glasses to correct eye defects.

c. 1280
Arab physician Ibn an-Nafis shows that blood flows through the lungs.

1300

1347–1350
Black Death (a bubonic plague pandemic) spreads through Europe, killing more than a quarter of its population.

1500

c. 1500
From his own dissections, Italian artist and scientist Leonardo da Vinci produces accurate anatomical drawings of the human body.

1543
Andreas Vesalius, a Flemish doctor, publishes *The Structure of the Human Body*, the first accurate description of human anatomy.

1545
Ambroise Paré, a French surgeon, publishes *Method of Treating Wounds* in which he describes his less painful, more successful techniques for treating wounds.

1561
Italian anatomist Gabrielle Fallopio (Fallopius) publishes *Anatomical Observations* in which he describes the duct linking the ovary to the uterus.

1562
Italian anatomist Bartolommeo Eustachio is the first to describe the ear in detail in his book *The Examination of the Organ of Hearing*.

1565
Swiss physician Paracelsus (Philipp von Hohenheim) publishes *Opus Chyrurgicum*, in which he attacks the works of Galen and Avicenna.

1590
Dutch instrument maker Zacharias Janssen invents the microscope.

1600

1603
Italian anatomist Hieronymus Fabricius publishes *On the Valves of Veins*, the first detailed description of vein structure.

1614
Italian physician Santorio Santorio (Sanctorius) publishes *The Art of Statistical Medicine*, the results of a 30-year study of his own bodily functions.

1628
William Harvey, an English doctor, publishes *On the Movement of the Heart*

and Blood, describing how blood circulates around the body, pumped by the heart.

1662

René Descartes' book *De homine*, published 12 years after his death, puts forward ideas about the brain and mind, and describes reflexes.

1663

Marcello Malpighi, an Italian physiologist and microscopist, discovers blood capillaries, helping to confirm that blood circulates around the body.

1664

Thomas Willis, an English doctor, describes the blood supply to the brain.

1665

English physicist Robert Hooke publishes *Micrographia*, in which he coins the term "cell".

1667

English physician Richard Lower carries out first blood transfusion to a human, using blood from a sheep.

1672

Regnier de Graaf, a Dutch doctor, describes the structure and workings of the female reproductive system.

1674–1677

Dutch draper and pioneer microscopist Antoni van Leeuwenhoek observes and describes red blood cells, sperm, and skeletal muscle cells using an early microscope.

1691

English doctor Clopton Havers makes the first description of the microscopic structure of bones.

1700

1717

Lady Mary Wortley Montagu, a British writer, brings the Turkish practice of smallpox inoculation to England.

1747

British naval doctor James Lind discovers that citrus fruits prevent the deficiency disease scurvy during long sea voyages.

1763–1793

British doctor John Hunter makes significant advances in knowledge about human anatomy, and elevates surgery from a craft to a science.

1775

French chemist Antoine Lavoisier discovers oxygen, and later shows that cell respiration is, like burning, a chemical process that consumes oxygen.

1780

Italian doctor Luigi Galvani experiments with nerves, muscles, and electricity.

1785

British doctor William Withering shows that the extracts of the foxglove plant could be used to treat heart failure.

1792

Austrian doctor Franz Gall begins his investigations into the link between behaviour and bumps on the skull. The investigations help to form the basis of his "science of phrenology".

1796

First vaccination against smallpox is carried out by British doctor Edward Jenner, when he takes pus from a cowpox blister and introduces it into the arm of an eight-year-old boy.

1800

1800

French doctor Marie François Bichat publishes *Traité des Membranes* in which he shows that organs are made of different groups of cells called "tissues".

1801

Philippe Pinel, a French doctor, suggests that mentally ill people should be treated more humanely.

1811

Scottish anatomist Charles Bell describes spinal nerve roots and shows that nerves are bundles of nerve cells.

1816

The stethoscope is invented by French doctor René Laënnec.

1817

English doctor James Parkinson first describes a brain disorder that affects movement in some older people, later to be called Parkinson's disease.

1818

British doctor James Blundell performs the first successful transfusion of human blood to a human patient.

1833

American surgeon William Beaumont publishes *Experiments and Observations on the Gastric Juice and the Physiology of Digestion*. The book records the results of his researches into the mechanism of digestion made on Alexis St. Martin, a man seriously wounded in a shooting accident.

1837

Czech biologist Johannes Purkinje first observes neurons (nerve cells) in the cerebellum of the brain, later called Purkinje cells.

1838

German scientists Theodor Schwann and Jakob Schleiden put forward their "cell theory", which states that all living things are made from cells.

1840

Jakob Henle, a German anatomist, states in his book *On Miasmas and Contagions* that infectious diseases are caused by micro-organisms.

1840

Charles Laveran, a French doctor, identifies the protist *Plasmodium* as the cause of the disease malaria.

1842

British surgeon William Bowman first describes the microscopic structure and function of the kidney.

1844

German doctor Carl Ludwig shows that nephrons in the kidneys act as filters for the production of urine.

1846

American dentist William Morton uses ether as an anaesthetic to make a patient unconscious and pain-free during an operation in Massachusetts General Hospital, USA.

1848

American railroad worker Phineas Gage survives an accident that drives an iron rod through the front of his brain, but suffers a behaviour change. This indicates to scientists that the frontal lobe of the cerebrum controls personality.

1848

Claude Bernard, a French scientist, demonstrates the function of the liver, and later shows that body cells need stable surroundings, thereby establishing the principles of what will later be called homeostasis.

1848

Hungarian doctor Ignaz Semmelweis demonstrates that hand washing by medical staff dramatically reduces deaths of women from puerperal (childbirth) fever.

1849

English-born American Elizabeth Blackwell becomes the first woman to qualify as a doctor in the USA.

1851

Hermann von Helmholtz, a German physicist, invents the ophthalmoscope (an instrument for looking inside the eye).

1854

British doctor John Snow halts an outbreak of cholera in London by removing the handle pump in Broad Street, suggesting that the disease is spread by contaminated water.

1858

In his book *Cellular Pathology*, German biologist Rudolf Virchow states that all cells are made from existing ones, and that diseases occur when cells stop working normally. This establishes the basis for the branch of medicine called pathology.

1859

Charles Darwin, a British scientist, puts forward the theory of evolution in his ground-breaking book *The Origin of Species*.

1860s

Louis Pasteur, a French scientist, explains how micro-organisms cause infectious diseases.

1861

Pierre Paul Broca, a French doctor, identifies the area of the brain (now Broca's area) that controls speech.

1865

British doctor Joseph Lister first uses carbolic acid as an antiseptic during surgery and dramatically reduces deaths from infection.

1870

Elizabeth Garrett Anderson begins practising as the first woman doctor in Britain.

1872

Italian doctor Camillo Golgi devises a stain that, for the first time, shows the brain's nerve cells clearly under the microscope.

1874

Carl Wernicke, an Austrian doctor, identifies the area on the left side of the brain (to be called Wernicke's area) that controls the understanding of spoken and written words.

1882

German doctor Robert Koch identifies the bacterium (*Mycobacterium tuberculosis*) that causes TB (tuberculosis).

1888

French microbiologists Emile Roux and Alexandre Yersin show that bacteria release toxins (poisons), which cause the symptoms of many diseases.

1889

Spanish physiologist Ramón Santiago y Cajal states that the nervous system is made up of a network of distinct nerve cells (later called neurons) that do not touch.

1895

X-rays are discovered by German physicist Wilhelm Roentgen.

1897

Ronald Ross, a British doctor, shows that the micro-organism (a protist) causing malaria is spread from person to person by *Anopheles* mosquitoes.

1900

Christiaan Eijkmann, a Dutch doctor, shows that the deficiency disease beriberi can be treated by a change in diet. This helps to establish the concept of "essential food factors", later called vitamins.

1900

Austrian doctor Sigmund Freud publishes *The Interpretation of Dreams*, which contains the basic ideas of psychoanalysis.

1900–1901

American army surgeon Walter Reed and his team demonstrate that yellow fever is transmitted by *Aëdes* mosquitoes and is caused by a virus.

1901

Austrian-American doctor Karl Landsteiner demonstrates the existence of blood groups (later classified as A, B, AB, and O), paving the way for safe blood transfusions. He was awarded the Nobel Prize for Medicine in 1930.

1901

Japanese biochemist Jokichi Takamine is the first scientist to isolate crystals of a pure hormone, adrenaline.

1902

British physiologists Ernest Starling and William Bayliss isolate secretin, the first substance to be named a hormone (a term that was later devised by Starling in 1905).

1903

An early version of the ECG (electrocardiograph), a device for monitoring heart activity, is invented by Dutch physiologist Willem Einthoven.

1905

Ernest Starling devises the term "hormone" to describe the newly discovered "chemical messengers" that co-ordinate body processes.

1906

Charles Sherrington, a British physiologist, publishes *The Integrative Action of the Nervous System*, a landmark work describing how the nervous system works.

1906–1912

Frederick Gowland Hopkins, a British biochemist, demonstrates the importance of "accessory food factors" (vitamins) in food.

1907

German neurologist Alois Alzheimer first describes the brain disorder (later to be called Alzheimer's disease) that causes a progressive decline in mental abilities.

1910

German scientist Paul Ehrlich discovers salvarsan, the first drug used to treat a specific disease.

1912

Polish-American biochemist Casimir Funk coins the term "vitamin" to describe essential nutrients needed in small amounts for normal body functioning.

1912

American Harvey Cushing publishes *The Pituitary Gland and its Disorders* in which he describes the functioning of the gland.

1914
American doctor Joseph Goldberger shows that pellagra is not an infectious disease but is caused by poor diet (later shown to be lack of the vitamin niacin).

1916
American birth control pioneer Margaret Sanger opens her first clinic in Brooklyn, USA.

1918
Edward Mellanby, a British scientist, discovers vitamin D, which is essential for normal bone growth.

1921
British birth control pioneer Marie Stopes opens her first clinic in London.

1921
Canadian physiologists Frederick Banting and Charles Best isolate the hormone insulin, enabling the disease diabetes to be controlled.

1921
German-born American scientist Otto Loewi detects chemicals called neurotransmitters involved in carrying signals between neurons.

1926
William Castle, an American doctor, demonstrates the intrinsic factor secreted by the stomach that aids the body's intake of vitamin B_{12}.

1928
Scottish bacteriologist Alexander Fleming discovers penicillin, the first antibiotic, when he notices mould growing on a plate of bacteria.

1928
Hungarian-born Albert von Szent-Györgyi, an American biochemist, isolates vitamin C.

1929
English physiologists Henry Dale and H.W. Dudley demonstrate the chemical transmission of nerve impulses between neurons, and identify acetylcholine as the first neurotransmitter.

1930
American physiologist Walter Cannon devises the term "homeostasis" – from the Greek for "standing still" – to describe the mechanisms whereby the body maintains a stable internal state.

1933
German electrical engineer Ernst Ruska invents the electron microscope.

1937
German-British biochemist Hans Krebs discovers the sequence of reactions called the Krebs cycle (or the citric acid cycle), which breaks down glucose during aerobic respiration to release energy.

1943
Dutch doctor Willem Kolff invents the kidney dialysis machine to treat people with kidney failure.

1948
World Health Organization (WHO) formed within the United Nations.

1952
British scientists Alan Hodgkin and Andrew Huxley describe nerve impulses.

1953
Using research by British physicist Rosalind Franklin, US biologist James Watson and British physicist Francis Crick discover the structure of DNA.

1953
American surgeon John Gibbon develops the heart-lung machine to pump blood during heart surgery.

1954
First use of polio vaccine developed by American doctor Jonas Salk.

1954
First successful kidney transplant carried out in Boston, USA.

1958
Ultrasound first used to check health of fetus in its mother's uterus by British professor Ian Donald.

1965
American biochemist Marshall Nirenberg finishes deciphering the genetic code through which DNA controls production of proteins inside a cell.

1967
South African surgeon Christiaan Barnard carries out first successful heart transplant.

1967
Introduction of mammography, an X-ray technique for detecting breast cancer.

1969
British biochemist Dorothy Hodgkin determines the structure of insulin using X-ray crystallography, having previously described that of penicillin in 1946 and vitamin B_{12} in 1956.

1970s
Discovery of natural painkillers, called endorphins and enkephalins, produced by the body.

1972
CT (computerized tomography) scanning first used to produce images of body organs.

1977
Last recorded case of smallpox; the disease is declared eradicated in 1979.

1978
Successful IVF (in vitro fertilization) by British doctors Patrick Steptoe and Robert Edwards results in first "test tube" baby, Louise Brown.

1980
Introduction of "keyhole" surgery, using an endoscope to look inside the body through small incisions.

1980s
PET scans first used to produce images of brain activity.

1981
AIDS (acquired immune deficiency syndrome) identified as a new disease.

1982
First artificial heart, invented by US scientist Robert Jarvik, implanted into a patient.

1984
Luc Montagnier, a French scientist, discovers the virus called HIV (human immunodeficiency virus), which causes AIDS.

1990
Human Genome Project is launched to identify the genes in human chromosomes.

1999
Chromosome 22 becomes the first human chromosome to have its DNA sequenced.

2000

2000
First "draft" of Human Genome Project completed.

2002
Gene therapy used to cure boys suffering from an inherited immunodeficiency disease which would otherwise leave the body defenceless against infection.

GLOSSARY

Abdomen
Lower part of the trunk (central part of the body) between the thorax (chest) and the hips.

Absorption
Process by which the products of digestion pass through the wall of the small intestine into the bloodstream.

Accommodation
Adjustment made by changing the shape of the lens of the eye so it can focus on near or distant objects.

Acne
Skin disorder causing spots that results from inflamed sebaceous glands and hair follicles.

Adolescence
Transition period between childhood and adulthood that occurs during the teenage years.

Aerobic respiration
Release of energy from glucose that takes place inside cells and requires oxygen.

Alimentary canal
Hollow tube which extends from the mouth to the anus, and includes the pharynx, oesophagus, stomach, and small and large intestines.

Allergy
Illness caused by over-reaction of the body's immune system to a normally harmless substance.

Alveoli (sing. Alveolus)
Microscopic air bags inside the lungs through which oxygen enters, and carbon dioxide leaves, the bloodstream.

Amino acid
One of a group of 20 chemical compounds that are the basic building blocks from which proteins are made.

Amputation
Surgical removal of all or part of an arm or leg.

Anaerobic respiration
Release of energy from glucose that takes place inside cells and does not use oxygen.

Anaesthetic
Drug used to temporarily abolish feelings of pain in a patient during surgery or while giving birth.

Anatomy
Study of the structure of the body, and how its parts relate to one another.

Angiogram
Special type of X-ray that reveals the outline of blood vessels after a dye that absorbs X-rays has been injected into them.

Antibody
Substance released by lymphocytes of the immune system that disables a pathogen and marks it for destruction.

Antigen
Foreign substance, usually found on the surface of pathogens such as bacteria, that triggers the immune system to respond.

Antiseptic
Chemical applied to the skin to destroy bacteria and other micro-organisms before they can cause infection.

Apgar score
System of scoring used to assess the condition of a newborn baby.

Appendicular skeleton
Part of the skeleton made up of the bones of the pectoral and pelvic girdles, and those of the upper and lower limbs.

Arteriole
Very small artery that delivers blood to a capillary.

Artery
Blood vessel that carries blood from the heart towards the tissues.

Association neuron
Neuron (nerve cell) that relays nerve impulses from one neuron to another, and processes information.

Atom
Smallest particle of an element, such as carbon or hydrogen, that can exist, and one of the building blocks from which all matter is made.

ATP (adenosine triphosphate)
Substance that stores, carries, and releases energy.

Atrium (pl. Atria)
Left or right upper chamber of the heart.

Autonomic nervous system (ANS)
Part of the nervous system that controls the involuntary activities of internal organs, such as heart rate and blood pressure.

Axial skeleton
Central part of the skeleton consisting of the skull, backbone, ribs, and sternum.

Axon
Also called a nerve fibre, this is the long "tail" of a neuron that carries nerve impulses away from its cell body.

Bacteria (sing. Bacterium)
Group of single-celled micro-organisms, commonly known as germs, some of which cause diseases such as typhoid and TB.

Base
One of four nitrogen-containing substances – adenine, cytosine, guanine, and thymine – that spell out the instructions written in genetic code in molecules of DNA.

Biopsy
Removal of a small piece of tissue from the body for examination under the microscope to look for signs of disease.

Blind spot
Also called the optic disc, this is the part of the retina of the eye where the optic nerve leaves the eye, and where light cannot be detected.

Blood vessel
Tube, such as an artery or vein, that carries blood through the body.

Body language
Form of non-verbal communication that uses body position, gestures, and facial expressions.

Brain stem
Lowest part of the brain. It is connected to the spinal cord, and controls vital functions such as breathing.

Cancer
One of a number of different diseases caused by body cells dividing out of control and producing growths called tumours.

Capillary
Microscopic blood vessel that carries blood from arterioles to venules and supplies individual cells.

Carbohydrate
One of a group of organic compounds, made up of carbon, hydrogen, and oxygen, that includes glucose and glycogen, and provides the body's main energy supply.

Carbon dioxide
Gas that is a waste product of cell respiration, which is released into the air from the lungs during exhalation (breathing out).

Cardiac muscle
Type of muscle found only in the heart.

Cartilage
Tough, flexible connective tissue that helps support the body and covers the ends of bones where they meet at joints.

Catalyst
Substance that speeds up the rate of a chemical reaction but does not change itself.

Cell body
Part of a neuron (nerve cell) that contains its nucleus.

Cell division
Process by which cells multiply by dividing into two.

Cell membrane
Also called a plasma membrane, this is the thin membrane that surrounds a cell and separates it from its environment.

Cell (Internal) respiration
The release of energy from glucose and other fuels that takes place inside cells.

Central nervous system (CNS)
The part of the nervous system that consists of the brain and spinal cord.

Cerebellum
Part of the brain that controls balance and ensures that movements are smooth and co-ordinated.

Cerebral cortex
Thin surface layer of the cerebrum that processes information relating to thought, memory, the senses, and movement.

Cerebrospinal fluid
Watery fluid that circulates within and around the central nervous system, and helps protect and nurture the brain and spinal cord.

Cerebrum
The largest part of the brain, which is involved in conscious thought, feelings, and movement.

Chemical digestion
The breakdown of food into simple molecules using enzymes.

Chemoreceptor
Receptor, such as those in the nose and tongue, that responds to chemicals dissolved in water.

Chromosome
One of 46 thread-like packages of DNA found inside most body cells that contain genes, the instructions needed to construct and run a body.

Chyme
Creamy, soup-like liquid containing semi-digested food that passes from the stomach into the small intestine during digestion.

Cilia (sing. Cilium)
Microscopic, hair-like projections from certain cells that beat in a rhythmic, wave-like manner to move things, such as mucus, across their surface.

Cochlea
Coiled structure inside each ear that detects sounds.

Collagen
Tough, fibrous protein that helps to strengthen cartilage, tendons, and other types of connective tissue.

Compact bone
Also called cortical bone, this is the very hard material that forms a bone's outer layer.

Computed tomography (CT)
Scanning technique that uses X-rays and computers to produce "slices" through living tissues.

Conception
The period between fertilization and the implantation of an embryo in the lining of the uterus.

Cone
One of two types of light receptors in the retina of the eye, cones provide colour vision and work in bright light.

Connective tissue
Tissue, such as bone or cartilage, that supports the body and holds together its various structures.

Consciousness
Awareness of self and surroundings produced by the cerebrum, which enables a person to make decisions and to know what they are doing.

Contagious disease
Infectious disease, such as the common cold or measles, that is easily passed from person to person.

Contraception
Use of various methods to prevent pregnancy.

Convex lens
Lens, such as the one found in the eye, that curves outwards on both surfaces, and which makes light rays converge (come together).

Cornea
Clear area at the front of the eye that allows light in and refracts (bends) light rays.

Cranial nerve
One of the 12 pairs of nerves that arise from the brain.

Cranium
Upper part of the skull, made from eight interlocking bones, that surrounds the brain.

Cytoplasm
Jelly-like fluid that fills a cell between the cell membrane and nucleus.

Deficiency disease
Disease caused by a lack of a particular nutrient in the diet, especially a vitamin or a mineral.

Dendrite
Short filament that carries nerve impulses to the cell body of a neuron.

Dentine
Hard, bone-like tissue that surrounds the pulp of a tooth, and gives the tooth its basic shape.

Dermis
Lower, thicker layer of the skin that contains blood vessels, sweat glands, and sensory receptors.

Diaphragm
Dome-shaped sheet of muscle that separates the thorax (chest) from the abdomen, and which plays a key role in breathing.

Diastole
Part of the heartbeat cycle when either the atria or ventricles are relaxed.

Diffusion
Random movement of molecules in a gas or liquid from an area of high concentration to one of low concentration until they are evenly distributed.

Digestion
Breakdown of complex molecules in food into simpler substances that can be absorbed into the bloodstream.

Diploid cell
One, like most body cells, that contains two sets of 23 chromosomes.

Dissacharide
Sugar, such as maltose, sucrose, or lactose, that is made up of two monosaccharide units.

DNA (Deoxy-ribonucleic acid)
One of a number of large molecules, each consisting of two intertwined nucleic acid strands, found inside body cells, which carry the genetic instructions needed to build and operate that cell.

Double helix
Name given to the twin strands of nucleic acids that spiral round each other like a twisted ladder in each DNA molecule.

Duct
A tube that leads from a gland and carries its products, such as the tear duct that carries tears from the tear glands.

Eardrum
Thin membrane at the end of the ear canal that vibrates when sounds hit it.

Egestion
Removal from the body in the form of faeces the undigested waste remaining after digestion.

Elastin
Protein whose fibres can stretch and recoil like a rubber band and give elasticity to connective tissues, such as those in the dermis of the skin.

Electrocardiogram (ECG)
Recording of the electrical changes that occur as the heart beats made by an electrocardiograph.

Electroencephalogram (EEG)
Recording of brain waves produced by electrical activity in the brain made by an electroencephalograph.

Electron microscope
Powerful microscope that uses an electron beam instead of light to produce highly magnified views of body cells and tissues.

Embryo
Name given to an unborn child during the first eight weeks of development after fertilization.

Enamel
Hardest material found inside the body, which covers the crown of a tooth.

Endocrine gland
Gland, such as the pituitary gland, that secretes hormones into the bloodstream.

Enzyme
Protein that acts as a biological catalyst to speed up the rate of chemical reactions both inside and outside cells.

Epidemic
Outbreak of an infectious disease that affects many people in the same location at the same time.

Epidermis
Upper, thinner, protective layer of the skin, from which the topmost layer of dead cells is constantly worn away and replaced from below.

Epithelium
Also called epithelial tissue, a sheet of cells, one or more cells thick that covers the body, lines its internal cavities, and forms glands.

Excretion
Elimination from the body of waste products produced by cell metabolism or of substances that have entered the bloodstream and are surplus to requirements.

Exhalation
The movement of air out of the lungs; also called expiration or breathing out.

Exocrine gland
Gland, such as a salivary or sweat gland, that secretes chemicals along a duct onto the body surface or into a body cavity.

Faeces
Solid waste consisting of undigested food, dead cells, and bacteria that remains after digestion and is eliminated from the body through the anus.

Fat
Type of lipid that is solid at room temperature, which is found in many foods.

Fatty acid
Building block, with glycerol, of fats and oils.

Feedback system
Control mechanism that maintains a stable state in the body by correcting unwanted changes and which regulates, for example, body temperature.

Fertility
Ability of a man and a woman to produce children without undue difficulties.

Fertilization
Joining together of an ovum and a sperm to make a new individual.

Fetus
Name given to the unborn child from the ninth week after fertilization until birth.

Fibre (dietary)
Also called roughage, this is indigestible plant material that gives bulk to food and improves the efficiency of intestinal muscles.

Fibre (muscle)
Name given to a muscle cell.

Follicle
Cluster of cells found inside the ovary that contains an ovum (see also Hair follicle).

Fracture
Break in a bone, often caused by a fall.

Fungi (sing. Fungus)
Group of living organisms, including yeasts and mushrooms, some of which are parasitic on humans, causing diseases such as athlete's foot.

Gas exchange
Movement of oxygen from the lungs into the bloodstream, and that of carbon dioxide in the opposite direction.

Gene
Carries the instructions to make a specific protein, and is one of the 30,000–50,000

genes stored in the DNA that makes up chromosomes inside a body cell.

Genetic code
Code used to convert the "message" carried by the sequence of bases in DNA into a sequence of amino acids to make a protein.

Genetic engineering
Artificial alteration made to the genetic make-up of an organism.

Genetics
Study of inheritance and transmission of genes from one generation to the next.

Girdle
Ring of bones that attaches the limbs to the rest of the skeleton.

Gland
Group of cells that produce chemical substances and release them into or onto the body.

Glial cells
Also called neuroglia, these cells protect and nurture neurons (nerve cells).

Glucose
Main sugar found circulating in the bloodstream, and the body's primary energy source.

Glycogen
Polysaccharide made of glucose subunits that forms a carbohydrate energy store in liver cells and muscle fibres.

Grey matter
Surface layer of the cerebrum, and inner part of the spinal cord, which consists mainly of neuron cell bodies.

Haemoglobin
Oxygen-carrying, iron-containing protein found inside red blood cells.

Hair follicle
Deep, hollow space in the skin from which a hair grows.

Haploid cell
A cell, such as a sperm or ovum, that is formed by meiosis and contains only a single set of 23 chromosomes.

Haversian system
Also called an osteon, this is a cylindrical collection of concentric bony tubes that is the basic units of compact bone.

Hepatic
Related to the liver, for example the hepatic artery that supplies blood to the liver.

Heredity
The passing on of characteristics controlled by genes from one generation to the next.

Homeostasis
Maintenance of stable conditions, including body temperature and blood glucose levels, regardless of external conditions.

Hormone
Chemical messenger that is produced and released by an endocrine gland and carried to its "target" by the blood.

Hunter-gatherer
Typical of early human societies but rarer today, person who lives by hunting animals and gathering plants rather than through agriculture.

Hypothalamus
Small but important part of the brain that regulates many body activities, including thirst and body temperature, by way of the pituitary gland and the autonomic nervous system.

Immune system
Collection of cells within the circulatory and lymphatic systems that protect the body from disease-causing micro-organisms.

Immunity
Ability of the immune system to "remember" and provide resistance to specific disease-causing micro-organisms.

Immunization
Provision of immunity against a disease by injecting a vaccine containing a weakened form of the micro-organism that causes that disease.

Infant mortality rate
Number of infants who die during the first year of life per 1,000 live births.

Infection
Establishment of disease-causing micro-organisms, such as bacteria, in the body.

Infectious disease
Disease, such as chickenpox, which is caused by a specific micro-organism.

Infertility
Inability of either a man or a woman, or both partners, to produce a child.

Ingestion
Taking food or drink into the body through the mouth.

Inhalation
The movement of air into the lungs; also called inspiration or breathing in.

Insertion
Attachment point of a muscle, through its tendon, to a bone that moves.

Insoluble
A substance that does not dissolve in water.

Integumentary system
External protective covering of the body provided by the skin, hair, and nails.

IVF (in vitro fertilization)
Technique used to help infertile couples conceive by fertilizing an ovum outside the body, then returning it to the uterus to develop.

Karyotype
Complete set of chromosomes inside a cell, photographed and arranged in pairs in descending size order.

Keratin
Tough, waterproof protein found inside cells making up hair, nails, and the upper epidermis of the skin.

Labour
Contractions of the muscular wall of the uterus before and during birth.

Ligament
Tough strips of fibrous connective tissue that hold bones together where they meet at joints.

Light microscope
Instrument that uses light rays focused by glass lenses to produce a magnified image of an object.

Limbic system
Part of the brain at the base of the cerebrum that controls emotions.

Lipid
One of a group of organic compounds, made up of carbon, hydrogen, and oxygen, that includes fats and oils (made of fatty acids and glycerol), phospholipids, and steroids, such as cholesterol.

Lymph
Fluid that flows through the lymphatic system from the tissues to the blood.

Lymphocyte
Type of white blood cell that plays a key role in the immune system.

Macronutrient
Nutrient – such as carbohydrate, fat, or protein – needed in large amounts by the body.

Macrophage
White blood cell present in a number of tissues that engulfs bacteria and foreign debris and plays a part in the immune system.

Magnetic resonance imaging (MRI)
Scanning technique that uses magnetism, radio waves, and a computer to produce images of the inside of the body.

Magnetoencephalo-graphy (MEG)
Scanning technique that produces real-time images of brain activity.

Mammal
Living organism that belongs to a group of animals that are "warm blooded", have a covering of fur, and feed their young with milk.

Marrow
Soft fatty tissue, either red or yellow, found in the spaces within bones.

Mechanical digestion
Breakdown of food into smaller particles through chewing or the churning action of stomach muscles.

Mechanoreceptor
Receptor that can detect pressure produced by touch, sound waves, or the stretching of muscles.

Medieval
Relating to the Middle Ages between the 5th and 15th centuries.

Meiosis
Type of cell division that occurs in the ovaries and testes to produce sex cells – ova and sperm – that contain a single set of chromosomes.

Melanin
Brown-black pigment found in the skin, hair, and the iris of the eye that gives them their colouring.

Membrane
Thin layer made up of epithelial tissue supported by connective tissue that covers or lines an external or internal body surface (see also Cell membrane).

Meninges (sing. Meninx)
Protective membranes that cover the brain and spinal cord.

Menopause
Period of a woman's life, between the ages of 45 and 55, when ovulation and menstrual periods cease.

Menstrual cycle
Sequence of changes, repeated about every 28 days, that prepares the lining of a woman's uterus to receive an ovum should it be fertilized.

Mesopotamia
Ancient region of south-western Asia between the Rivers Tigris and Euphrates, known today as Iraq.

Metabolism
Sum of all the chemical processes that take place within the body, particularly within its cells.

Metabolic rate
Rate at which energy is released by metabolism.

Microbe
General name for a micro-organism that causes disease.

Micrograph
Photograph taken with the aid of a microscope.

Micronutrient
Nutrient – such as a vitamin or mineral – needed in small amounts by the body.

Micro-organism
Tiny organism, such as a bacterium, that can only be seen with a microscope.

Middle Ages
Period of Western European history between the 5th and 15th centuries.

Mineral
One of about 20 chemical elements, including calcium and iron, that must be present in the diet to maintain good health.

Mitochondria (sing. Mitochondrion)
Organelles inside cells that carry out aerobic respiration to release energy.

Mitosis
Type of cell division used for growth and repair that produces two identical cells from each "parent" cell.

Molecule
Chemical unit that is made up of two or more linked atoms, such as the two hydrogen atoms and one oxygen atom in a water molecule.

Monosaccharide
Sugar such as glucose, fructose, or galactose, that is the simplest type of carbohydrate.

Motor neuron
Neuron (nerve cell) that carries nerve impulses from the central nervous system to muscles and glands.

Mucous membrane
Layer that lines the body cavities which open to the exterior, for example the respiratory system, and which secretes mucus.

Mucus
Thick, slimy fluid secreted by mucous membranes, which moistens, protects, and lubricates.

Muscle tone
Partial contraction of a muscle that maintains the body's posture.

Mutation
Change to the DNA in one of a cell's chromosomes that may have harmful effects, and can be passed on to the next generation.

Myelin sheath
Insulating sheath wrapped around most axons (nerve fibres) that increases the speed of conduction of nerve impulses.

Myofibril
One of thousands of tiny rod-like strands inside a muscle fibre (cell).

Nephron
One of a million filtration units inside each kidney that produce urine.

Nerve
Cable-like bundle of neurons (nerve cells) that relays nerve impulses between the body and central nervous system.

Nerve fibre
Also called an axon, this is the long "tail" of a neuron that carries nerve impulses away from its cell body.

Nerve impulse
Tiny electrical signal that passes along a neuron (nerve cell) at high speed.

Neuron
One of the billions of interconnected nerve cells that carries electrical signals at high speed and makes up the nervous system; the brain, spinal cord, and nerves.

Neurotransmitter
Chemical released when a nerve impulse reaches the end of a neuron, which triggers a nerve impulse in a neighbouring neuron.

Nitrogen
Gas, like oxygen, that is found in the air but which normally plays no part in body functions.

Nobel Prize
Prestigious award given annually for outstanding achievement in one of five fields, including physiology or medicine.

Non-infectious disease
Disease, such as cancer or heart disease, that is not caused by a disease-causing micro-organism.

Nucleic acid
Organic compound, such as DNA or RNA, which contains carbon, hydrogen, oxygen, nitrogen, and phosphorus, and is made up of units called nucleotides.

Nucleotide
Basic building block of nucleic acids such as DNA, consisting of a phosphate group, a deoxyribose sugar, and a nitrogenous base (adenine, cytosine, guanine, or thymine).

Nucleus
Control centre of a cell that contains chromosomes.

Nutrient
Substance – such as carbohydrate, protein, fat, vitamin, or mineral – needed in the diet to maintain good health and normal body functioning.

Obstetrician
Doctor who specializes in pregnancy and childbirth.

Ophthalmoscope
Instrument used to view the inside of the eye.

Orbit
Socket in the skull that surrounds, supports, and protects the eyeball.

Organ
Body part, such as the kidney or brain, with a specific role or roles that is made up of two or more different types of tissues.

Organic compound
Substance, such as carbohydrate, protein, lipid, or nucleic acid, that has a carbon "skeleton", and is made only by living systems.

Organelle
Microscopic structure inside a cell, such as a mitochondrion, that has a specific function.

Origin
Attachment point of a muscle, through its tendon, to a bone that is stationary.

Osmoregulation
Maintenance of the correct levels of water and salts in blood and tissue fluids carried out by the kidneys.

Ossicle
One of the three small, sound-transmitting bones found inside the middle ear.

Ossification
Process of bone formation.

Ovarian cycle
Sequence of changes, repeated about every 28 days, that causes an ovum to be released from a woman's ovary.

Ovulation
Release of an ovum from a woman's ovary.

Ovum (pl. Ova)
Also called an egg, this is the female sex cell, which is produced by, and released from, a woman's ovary.

Oxygen
Gas found in the air which is taken into the bloodstream through the lungs and used by body cells in aerobic respiration to release energy from glucose.

Palaeontologist
Scientist who studies fossils.

Papillae (sing. Papilla)
Small bumps projecting from the tongue's surface, some of which house taste receptors called taste buds.

Paraplegic
Weakness or paralysis of both legs and sometimes part of the trunk caused by damage to the spinal cord.

Pathogen
Disease-causing micro-organism such as a bacterium, virus, protist, or fungus.

Pectoral girdle
Girdle formed by the two collar bones and two shoulder blades, which attaches the arms to the skeleton.

Pelvic girdle
Girdle formed by the two hip bones which anchors the legs to the skeleton and, with the sacrum, forms the basin-like pelvis.

Periosteum
Membrane covering the surface of bones, which contains blood vessels.

Peripheral nervous system (PNS)
The part of the nervous system that consists of the nerves that relay nerve impulses between the body and the central nervous system.

Peristalsis
Wave of muscular contraction through a hollow organ that, for example, pushes food down the oesophagus or urine down a ureter.

Phagocyte
General name for white blood cells – including neutrophils and macrophages – that engulf and digest disease-causing micro-organisms.

Phagocytosis
Process by which phagocytes engulf and digest disease-causing micro-organisms.

Phantom pain
Sensation of pain felt in a limb that is no longer present because it has been amputated.

Phospholipid
Type of phosphate-containing lipid that makes up the cell membranes around cells, and the membranes around organelles.

Physiology
Study of how the body works and functions.

Pitch
Quality of a sound – whether high- or low-pitched – that depends on the frequency of sound waves, that is, how quickly one wave is followed by the next.

Placenta
Organ that develops in the uterus during pregnancy that forms an interface between the blood supplies of the mother and fetus, through which the fetus receives food and oxygen.

Plasma
Liquid part of the blood, which is mainly water.

Polysaccharide

Complex carbohydrate, such as glycogen, that does not have a sweet taste and is made up of long chains of monosaccharides, such as glucose.

Portal system

Veins that carry blood from one organ to another rather than towards the heart.

Positron emission tomography (PET)

Scanning technique that uses radioactive substances injected into the body to show parts of the body at work, especially the brain.

Pregnancy

The period from conception to birth, but which is dated from the start of a woman's last menstrual cycle, and so lasts about 40 weeks.

Primates

Group of mammals which includes monkeys, apes, and humans.

Protein

One of a group of organic compounds, made up of carbon, hydrogen, oxygen, nitrogen, and sulphur, that perform many roles inside the body including making enzymes.

Protists

Group of single-celled organisms, some of which cause diseases in humans such as malaria.

Puberty

Period during adolescence when the body grows and develops an adult appearance, and the reproductive system starts working.

Pulmonary circulation

Part of the circulatory system that carries blood from the heart to the lungs and back to the heart.

Pupil

Opening in the centre of the iris through which light enters the eye.

Radionuclide scanning

Scanning technique that uses radioactive substances to reveal the functioning of organs such as bones.

Receptor

Special cells or neurons that detect stimuli, such as light, and trigger sensory neurons.

Reflex

Automatic, unconscious, split-second response to a stimulus that often protects the body from danger.

Reflex arc

Nervous pathway, often through the spinal cord but not the brain, involved in a reflex.

Renaissance

Term meaning "rebirth", which describes the period between the 14th and early 17th centuries in Europe when there was a creative revolution in the arts, sciences, and medicine.

Renal

Related to the kidney, as in the renal artery that supplies blood to the kidney.

Retina

Inner lining of the eyeball that is packed with light receptors.

Rod

One of two types of light receptors in the retina of the eye, rods provide black and white vision, and work best in dim light.

Saliva

Fluid released into the mouth, especially during chewing, by the salivary glands.

Scanning electron micrograph (SEM)

Photograph produced using a scanning electron microscope.

Sebaceous gland

Gland connected to a hair follicle that produces an oily liquid called sebum.

Sebum

Oily liquid that keeps hair and skin soft, flexible, and waterproof.

Secretion

Chemical substance made and released by a gland.

Semen

Fluid produced by male reproductive glands, which activates and nurtures sperm, and in which they swim.

Semicircular canal

Part of the inner ear that is involved in balance.

Sensory neuron

Neuron (nerve cell) that carries nerve impulses from sensory receptors to the central nervous system.

Septum

Dividing wall within body parts, such as in the nose.

Skeletal muscle

Type of muscle that is attached to the skeleton and moves the body.

Smooth muscle

Type of muscle found inside the walls of organs that, for example, pushes food along the small intestine.

Soluble

Describes a substance that dissolves in water.

Solution

Mixture of one substance (called a solute) dissolved in another (the solvent); glucose and carbon dioxide are both solutes that dissolve in the water (solvent) in blood.

Species

A group of living things in the natural world that can breed with each other.

Sperm

Also called spermatozoa, these are the male sex cells, which are made in and released from a man's testes.

Spermatogenesis

Process of sperm production inside the testes.

Sphincter

Ring of muscle around a passage or opening that opens or closes to control the flow of, for example, food or urine along it.

Spinal cord

Column of nervous tissue that runs down the back and relays nerve signals between the brain and body.

Spinal nerve

One of the 31 pairs of nerves that arise from the spinal cord.

Spongy bone

Also called cancellous bone, this is the tough but lightweight honeycomb of struts and cavities that forms the inner part of a bone.

Stain

Dye used to colour cells and tissues so they can be seen under the light microscope.

Sterilization

Surgical procedure that stops a person being able to reproduce. Also the process used to destroy any micro-organisms on surgical instruments and other materials used in hospitals to minimize the risk of infection to patients.

ACKNOWLEDGEMENTS

The team at Dorling Kindersley would like to thank:
Carole Oliver, Clair Watson and Willie Wood for design help; Caryn Jenner and Brad Round for editorial help; Sophie Young for DTP help; Alyson Lacewing for proofreading; and Marie Lorimer for the index.

The publisher would like to thank the following for their kind permission to reproduce their photographs:
Key: a=above; b=below; c=centre; l=left; r=right; t=top; bi=background image; mi=main image

10 Telegraph Colour Library/Getty Images: Alistair Berg bl; 10–11 Science Photo Library: mi, l; Simon Fraser c; r Professors P. M. Motta, S. Makabe & T. Naguro c; 12–13 Harry Cutting mi; 14–15 Liam Bailey bl; source unknown cl; 16–17 Science Photo Library: Professor P. Motta, mi (circular), bl, tr (circular), Astrid & Hanns-Frieder Michler tr (cutout), Pascal Goetgheluck cb; Science Museum ct (cutout); Ann Ronan Picture Library: ct (circular); 18–19 Science Photo Library: mi, c (circular), br, Eric Grave tl, Quest bl, Professor P. Motta cl (circular); Mary Evans: tr; 20–21 Denis Models mi; Science Photo Library: Alfred Paskia l; Professors P. M. Motta, S. Makabe & T. Naguro tr, CNRI cr (circular), Dr Kari Lounatmaa br; 22–23 Science Photo Library: bi, Dr Arther Tucker cl, Professor K. Seddon & Dr T. Evans cr, Ken Eward br; 20–21 Denis Models mi; Science Photo Library: Alfred Paskia l, Professors P. M. Motta, S. Makabe & T. Naguro tr, CNRI cr (circular); Dr Kari Lounatmaa br; 22–23 Science Photo Library: bi, DR Arther Tucker cl, Professor K. Seddon & Dr T. Evans cr, Ken Eward br; 24–25 Science Photo Library: bl; A. Barrington Brown ct, Dr Gopal Murti tr; 26–27 Science Photo Library: Quest mi and ct, David M. Martin cl (circular), BSIP Laurent/H. Amercain cb (box), Professor P. Motta rt, Manfred Kage rc, cr; 28–29 Science Photo Library: Simon Fraser mi; 30–31 Science Photo Library: Volker Steger mi, SNRI tl, GJLP c, Dr John Mazziotta et al/Neurology rt, Alfred Paskia tr (cutout), cl, Simon Fraser/Royal Victoria Infirmary, Newcastle-upon-Tyne br; 32–33 Science Photo Library: Professor P. Motta mi, all other images Science Photo Library; 34–35 Geoff Brightling l and tr (cutout); Corbis Images ct; Science Photo Library: Martin Dohrn c (box), Andrew Syred tr (circular); Andy Crawford cb (cutout and circular); Dave King br (cutout) and (circular); 36–37 Science Photo Library: Dr Jeremy Burgess mi, Christina Pedrazzini c, Dr P. Marazzi cr, D. Phillips tr; Richard Wehr/Custom Medical Stock Photo/SPL cl; 38–39 Science Photo Library: bl, Dr P. Marazzi ct, Quest c, Geoff Brightling cb; Corbis Images: tr; Anthony Duke, digital artwork br; 40–41 Science Photo Library: D. Phillips mi; Andrew Syred bl and cr, Eye of Science r and br; Andy Crawford ct photograph (digitally amended by Peter Bull); 42–43 Science Photo Library: b (box), Professor P. Motta bi,

Biophoto Associates tr, CNRI bl, GJLP/CNRI bl; Mary Evans c (box); Andy Crawford mi; 44–45 Philip Dowell tl; Science Photo Library: Alfred Paskieka c (oval), Dave Roberts tr, Dave King br; 46–47 Science Photo Library: r, Allsport c, GJLP/CNRI tr (circular); 48–49 Science Photo Library: Department of Clinical Radiology, Salibury District Hospital bi, CNRI tl; Mehau Kulyk cb, tr; Ann Ronan Picture Library c (box); Philip Dowell r; 50–51 Science Photo Library: Hugh Turvey tl and ct, Mehau Kulyk bl; Andy Crawford c (x2); 52–53 Science Photo Library: bi; sources unknown bl and tr; Corbis Images: br; 54–55 Science Photo Library: Andrew Syred l, c and r, Professor P. Motta ct, Robert Becker/Custom Medical Stock Photo br; 56–57 Science Photo Library: Department of Clinical Radiology, Salibury District Hospital bi, bl, and tr, Quest cl, Scott Camazine ct; Mary Evans: bc (box); 58–59 Ray Moller cb; Dave King tr (cutout); Science Photo Library: Dr P. Marazzi and Department of Clinical Radiology, Salibury District Hospital (digitally amended by Anthony Duke) tr, Mike Devlin br (circular), Brad Nelson/Medical Stock Photo br; 60–61 Andy Crawford: mi photography, tc (digital artwork Peter Bull), and r; 6 artworks surrounding skeleton by Colin Salmon; 62–63 Science Photo Library: cl and c, Dr R. Clark & M. Goff br; M.I. Walker ct and br (box); 64–65 Andy Crawford c circular images (x2); Raymond Evans/Unit of Art in Medicine, University of Manchester, London r; 66–67 Science Photo Library: bl and rc; Andy Crawford tr; 68–69 The Royal Collection Picture Library bl (and bi); Wellcome Library, London: tl and cb; Science Photo Library: c; 70–71 Corbis c ; Getty Images cb; 72–73 Getty Images: mi; Corbis Images: bl; Quest c (circular); Andy Crawford and Steve Gordon tr; Science Photo Library: Will McIntyre br; 74–75 Science Photo Library: GJLP/CNRI r; all other images Science Photo Library; 76–77 Getty Images: mi; Science Photo Library: BSIP VEM tl, bl (box) and c (circular); Andy Crawford br photography (head and hand); brain artwork, eye and heart DK; 78–79 Science Photo Library: Quest bl, Nancy Kedersha ct, BSIP, Sercomi cb (box); Corbis Images tr; 80–81 Peter Bull br; Science Photo Library: Eye of Science ct (circular), tr and br; Frank Greenaway cr; 82–83 Science Photo Library: Quest bl, Sue Ford tr; 84–85 Andy Crawford mi photography; Getty Imags br; 86–87 Andy Crawford l photography (superimposed image DK); Corbis: ct; Andy Crawford cb; Science Photo Library: CC Studio tr; Mary Evans: br; 88–89 Science Photo Library: bl, clt, and crt (box); Andy Crawford r photography (superimposed image DK); 90–91 Imperial War Museum: crb (and bi); Department of Neuorology and Image Analysis Facility, University of Iowa cl; Andy Crawford ct; Science Photo Library: bl; Hank Morgon tr, BSIP Astier br; 92–93 Science Photo Library: GJLP/CNRI bi, ct, Getty Images: tr; 94–95 Steve Gorton bl; Corbis Images: c; Science Photo Library: tr; Dave

King br (x3); 96–97 Andy Crawford bl photography (superimposed image DK); Science Photo Library: Hank Morgan ct, BSIP Buntschucr cr; 98–99 Andy Crawford mi photography (superimposed artwork DK); Science Photo Library: Jean-Loup Charmet cr; 100–101 Geoff Dann and Andy Crawford l; Andy Crawford ct, c, and cb; Science Photo Library: Montreal Neuro Institute, McGill University/ CNRI cr; 102–103 Science Photo Library: Mehau Kulyk, bi, US National Library of Medicine cl, Simon Fraser (RNC Newcastle-upon-Tyne) cr; Mary Evans: cb; Science Museum: tr; 104–105 Andy Crawford mi photography (digitally amended by Anthony Duke); Corbis: bl; The Bridgeman Art Library: tr (box); Corbis Images: cr; Science Photo Library: BSIP Leca br; 106–107 Andy Crawford bi and l photography (superimposed artwork DK); Getty Images: tr; Science Photo Library: cb and br; 108–109 Andy Crawford tl photography (superimposed artwork DK); Science Photo Library: Quest cl, James King-Holmes cr (circular) and tr; Corbis: br; 110–111 Getty Images: mi; Andy Crawford tl photography (superimposed artwork DK); Science Photo Library: Mark Clarke br; 112–113 Andy Crawford mi photography (superimposed artwork Peter Bull) and br; Science Photo Library: cl and cr; 114–115 Mary Evans bl; Science Photo Library: PIR-CNRI tl, Bill Longcore tr, cbr Mark Burnett cbr; Custom Medical Stock Photo br; 116–117 Science Photo Library: Mehau Kulyk br (plus superimposed artwork DK); Ronald James cl; 118–119 Getty Images: mi; Science Photo Library: bl and tr; 120–21 Andy Crawford cl (superimposed artwork DK); Getty Images: tr; Science Photo Library: br Pascal Goetgheluk; 122–23 Andy Crawford mi photography (superimposed artwork Anthony Duke); Science Photo Library: bl; 124–25 Science Photo Library: Astrid & Hanns-Frieder Michler bi, tl, Saturn Stills c, J.C. Revy cr, Volker Steger bl; Banting House National Historic Site tr; 126–27 Science Photo Library: Quest mi, Professor P. Motta l, c, CNRI r; 128–29 Science Photo Library: National Cancer Institute bi; Simon Fraser tr, Quest bl; 130–31 Science Photo Library: Quest bl, tl, St Bartholomew's Hospital ct, Robert Becker, Custom Medical Photo Stock br; Andy Crawford cbr; 132–33 Science Photo Library: bi, CNRI br (circular), Francis Leroy, Bicosmos tr, Eye of Sciencebr; 134–35 Science Photo Library: Jurgen Berger mi, Max Planc Institute, Nigel Dennis bl; CNRI tr (circular) and br, Dr Gopal Murti tr; Quest cr, NIBSC br (circular), Secchi-Lecaque bcr; 136–37 Science Photo Library: bl and tr, NIBSC bcl, tl Volker Steger; Mary Evans: bl; 138–39 Corbis: br; Mary Evans: ct and cb; Science Photo Library: David Scharf bi; Jerry Mason br; 140–41 Science Photo Library: bl and tr, BSIP, Carvallini James ct, Quest cb, Phillipe Plailly br; 142–43 Science Photo Library: Phillipe Plailly ctr, bc; Getty Images: br; 144–45 Science Photo Library: CNRI bi and bl, Will & Denis

Mcintyre tr, D. Ouellette c, Ouellete & Theroux bl; Corbis: ctl ; **146–47** Science Photo Library: CNRI bi; **148–49** Science Photo Library: cl, ct, and tr (digitally manipulated by Anthony Duke); Mary Evans: br; **150–51** Science Photo Library: Saturn Stills tr; Corbis: cbr and br.; **152–53** Science Photo Library: tl John Griem; **154–55** Science Photo Library: Secchi Lecaque bi, Alferd Pasieka ctl, tl, Dr P. Marazzi cr; **156–57** Science Photo Library: Dr Linda Stannard tl; Eye of Science bl and ctr, Dr P. Marazzi bcl, Dr Gary Gaugler c, Quest tr; Mary Evans br; **158–59** Andy Crawford cl phography (superimposed image DK); Science Photo Library: circular images, clockwise from top: Manfred Kage, Professor P. Motta, Science Photo Library, Biophoto Associates, Professors P.M. Motta, K.R. Porter, and P.M. Andrews, Science Photo Library, Philip A. Harington tr, Juergen Bergercr and bi, Oscar Burriel r; **160–61** Science Photo Library: cl, Dr Andrejs Liepins tr, Jerrican br; **162–63** Science Photo Library: tl, br, c, cb, and bi; **164–65** Science Photo Library: Eye of Science br, James King-Holmes tl, cbr, ctr, and br, J. F. Wilson cr; Corbis: tr; **166–67** Science Photo Library: BSP Vem tl; bl, cl, and cr; **168–69** Science Photo Library: Clinique Ste Catherine, CNRI tl, tr, Alain Dex ctr ; Matt Meadows br, H. Schleichkorn br; **170–71** Science Photo Library: CNRI bi; SPL tr; Getty Images: br; **172–73** Andy Crawford bl photography (superimposed images); Science Photo Library: br; **174–75** Mary Evans: tl; Ann RonanPicture Library: bl; Science Photo Library: National Library of Medicine ct, Bill Longcore br; Corbis: bi; Mary Evans; **176–77** Science Photo Library: Mehau Kulyk l, Professor C. Ferland bcl and bcr; Labat tr; Andy Crawford c (x2) photography (superimposed images (x2) Peter Bull); cr (x2) Andy Crawford; Getty Images: br; **178–79** Andy Crawford c photography (superimposed images DK); Science Photo Library: Eye of Science br; **180–81** Andy Crawford mi; Science Photo Library: BSIP Vem tcr, CNRI cr, Dr Tony Brain br; **182–83** Science Photo Library: Biophoto Associates tr; Andy Crawford: bi, l photography (superimposed imges DK); Dr Klaus Schiller cl; **184–85** Science Photo Library: bl and bcr, Eye of Science bcl, Dr K. R. Schiller tr; Corbis: br; **186–87** Science Photo Library: Manfred Cage cl, bl and br, Professor P. Motta tcr; Andy Crawford mi photography (superimposed images DK); **188–89** Science Photo Library: Eye of Science cl, David M. Martin cl, Professor P. Motta tr; Andy Crawford mi and ctr photography (superimposed images DK); **190–91** Science Photo Library: cr, Professors P. Motta & F.M. Magliocca br; **192–93** Andy Crawford bi; Science Photo Library: Adam Hart-Davis far bl, Dr Jeremy Burgess bl, c; Corbis: tcl and tr; Getty Images: br; **194–95** All images DK; **196–97** Science Photo Library: Alfred Pasieka tl, (superimposed image DK), Dr Jeremy Burgess ctr **198–99** Andy Crawford tl photography (superimposed image DK); bl, c, tr and br DK; Professor P. Motta cr, bl; **200–01** Corbis: mi; Andy Crawford bl photography (superimposed image DK); **202–03** Science Photo Library: Brian Yarvin cr, br; **204–5** Science Photo Library: Quest mi, Brad Nelson bl, Biophoto Associates tl, Professors P. Motta and M.

Castellucci br (circular); Corbis: bl **206–07** Janos Marffy l; Science Photo Library: BSIP Vem cl, Mauro Fermariello tr, Andrew Syred br (circular); r DK; Andy Crawford bl photography, (superimposed image DK); br Corbis; **208–09** Science Photo Library: Quest bl, Ed Reschke, Peter Arnold Inc tl, tr, CNRI b; James Stevenson br; **210** Science Photo Library: BSIP VEM br; Professor P. Motta, G Macchiarelli, SA Nottola bl; Petit Format– Nestle bc; **210–211** Science Photo Library: TEK Image –211; **212** The Image Bank–Getty Images: Larry Dale Gordon br; **213** Science Photo Library: D. Phillips tr; Science Pictures Ltd br; **215** Corbis: Bohemian Nomad Picturemakers tr; Science Photo Library: Ed Reschke, Peter Arnold Inc. tl; National Library of Medicine br; The Wellcome Institute Library, London: Dr Joyce Harper bc; **216** Science Photo Library: CNRI bc; Astrid & Hanns-Frieder Michler ac; **216-217** Science Photo Library: Science Source tl; **217** Corbis: tr; Science Photo Library: Peter Cull tc; Custom Medical Stock Photo br; **218** Science Photo Library: br; Professor P. Motta, G. Macchiarelli, S. A. Nottola cl; **219** Science Photo Library: CNRI br;Custom Medical Stock Photo tr; Dr Yorgos Nikas tl; **221** Science Photo Library: Professor P. M. Motta & E. Vizza tr; Professors P. M. Motta & J. Van Blerkom tl; **223** Science Photo Library: Dr Yorgos Nikas tl, cla, clb; The Wellcome Institute Library, London: Yorgas Nikas br; **224** Corbis: Bettmann tl; **224–225** Science Photo Library: Pascal Goetgheluck c; **225** Corbis: Bettmann tr; Science Photo Library: Pascal Goetgheluck br; **226** Science Photo Library: Petit Format– Nestle crb; **227** Science Photo Library: Custom Medical Stock Photo br; **228** Science Photo Library: Dept of Clinical Radiology, Salisbury District Hospital crb; **229** Science Photo Library: Petit Format– Prof E. Symonds br; **231** The Wellcome Institute Library, London: l; **232** Corbis: Dick Clintsman cla; Science Photo Library: tr; Jean–Loup Charmet b; **233** Science Photo Library: Lawrence Migdale cr; Alfred Pasieka tr; **234** Corbis: Reed Kaestner br; Barnabas Kindersley: tr; Guy Ryecart cl; **235** Barnabas Kindersley: c, cr, bc; Guy Ryecart: c; **236** Science Photo Library: Christopher Briscoe tr; Ron Sutherland cl; **237** Telegraph Colour Library–Getty Images: Stephanie Rausser br; **238** Telegraph Colour Library–Getty Images: Jim Cummins b; **239** Corbis: O'Brien Productions tl; Science Photo Library: BSIP VEM clb; Mehau Kulyk br; **240** Science Photo Library: Scott Camazine tl; Quest c; **241** The Image Bank–Getty Images: Ghislain & Marie David de Lossy tr; Science Photo Library: Dr P. Marazzi c; Telegraph Colour Library–Getty Images: tl; **242** Science Photo Library: Dr P. Marazzi r; Telegraph Colour Library–Getty Images: Larry Bray cl; **243** Science Photo Library: Custom Medical Stock Photo r; Dr P. Marazzi tl; Alfred Pasieka cl; **244** Science Photo Library: CNRI cl, crb, bc; Mike Bluestone tr; **245** Pa Photos: bl; **246** Science Photo Library: Peter Menzel tl; **246–247** Science Photo Library: Biophoto Associates; **247** Science Photo Library: tr; **249** Science Photo Library: CNRI bc; Lauren Shear br; **250** Science Photo Library: Andrew Syred bl; **250–251** Science Photo Library: TEK Image; **251** Science Photo Library: BSIP, PIKO

tl; BSIP VEM cr; Klaus Guldbrandsen bl; **252** Bridgeman Art Library, London – New York: Bibliotheque des Arts Decoratifs, Paris, France; Science Museum: bc; Science Photo Library: National Library of Medicine bl; **254–255** American Museum Of Natural History: Denis Finnin & Craig Chesek b; **255** Corbis: Kevin Schafer br; Science Museum: tr; Werner Forman Archive: Tanzania National Museum, Dar es Salaam cr; **256** AKG London: Erich Lessing tr; © Michael Holford: cla; Science Photo Library: Christian Jegou, Publiphoto Diffusion br; **257** Science Museum: br; Charles Walker Collection: tl; **258** AKG London: tr; Erich Lessing bc; Charles Walker Collection: c; **258–259** Scala Group S.p.A.: **259** Mary Evans Picture Library: tr; World Health Organisation: cra; **260** Mary Evans Picture Library: tl; Science Museum: tr; Science Photo Library: National Library of Medicine bl; **260–261** AKG London: b; **261** Museum of the Royal Pharmaceutical Society: cla; Hulton Getty Archive: cr; **262** Science Photo Library: Jean-Loup Charmet tl; **262–263** AKG London: b; **263** Corbis: © Bettmann br; Science Photo Library: John Burbidge tl; CAMR–A B Dowsett tr; **264** AKG London: tl; Erich Lessing bc; Bridgeman Art Library, London – New York: Bibliotheque des Arts Decoratifs, Paris, France cr; **265** Science Museum: tl; Mary Evans Picture Library: br **266** Corbis: Bettmann tr; Science Photo Library: cl; **266–267** Science Museum: **267** AKG London: Erich Lessing br; Science Museum: tr; **268** Hulton Getty Archive: bc; Science Photo Library: Sheila Terry tl; **268–269** Science Museum: **269** AKG London: tr; Science Museum: crb; Hulton Getty Archive: bc; **270** AKG London: Musee d'Orsay br; The Image Bank–Getty Images: Roine Magnusson bl; Science Photo Library: Dr Gopal Murti tl; **271** AKG London: br; Mary Evans Picture Library: tl, ac; **272** AKG London: tl; Hulton Getty Archive: br; **273** Science Museum: br; Hulton Getty Archive: c; Sean Hunter: tc; Science Photo Library: tl; **274** Mary Evans Picture Library: tl; **274–275** Science Photo Library: Antonia Reeve b; **275** Science Photo Library: Martin Dohrn tr; James King-Holmes br; **276** The Image Bank–Getty Images: Wayne H. Chasan c; **277** Corbis: Lindsay Hebberd tc; **278** Science Photo Library: Simon Fraser tl; **278–279** Telegraph Colour Library–Getty Images: VCL–Paul Viant b; **279** Science Photo Library: CC Studio tc; Telegraph Colour Library–Getty Images: Bill Losh tr; **280** Science Photo Library: Custom Medical Stock Photo br; **282–83** Science Photo Library: CNRI mi, Quest l, cl; J. Croyle/Custom Medical Stockc; **288–289** Science Photo Library: Nancy Kedershami; **294–95** Science Photo Library: Professor P. Motta mi; **290–291** Science Photo Library: Astrid & Hanns-Frieder Michler; **292–293** Science Photo Library: Dr Gopal Murti;

Endpapers Science Photo Library: Juergen Berger, Max-Planck Institute.

Jacket: The Image Bank/Getty Images: Fer front c; Science Photo Library: back tl, back tcl; CNRI front tl; Mehau Kulyk front tr, back tr; Dr. Yorgos Nikas front tll; Alfred Pasieka front tc; D. Phillips back tcr.

All other images © Dorling Kindersley. For further information see: www.dkimages.com